METAPHOR AND KNOWLEDGE

SUNY series, Studies in Scientific and Technical Communication
James P. Zappen, editor

METAPHOR AND KNOWLEDGE

The Challenges of Writing Science

Ken Baake

State University of New York Press

Published by
State University of New York Press, Albany

For information, address State University of New York Press,
90 State Street, Suite 700, Albany, NY 12207

Production by Judith Block
Marketing by Jennifer Giovani

Library of Congress Cataloging-in-Publication Data
Baake, Ken
   Metaphor and knowledge : the challenges of writing science / Ken Baake.
      p. cm. — (SUNY series, Studies in scientific and technical communication)
   ISBN 0–7914–5743–5 (alk. paper) — ISBN 0–7914–5744–3 (ppk. : alk. paper)
   1. Technical writing. I. Title. II. Series

T11 .B23 2003
808′.0666—dc 21

                                                                                      2002030479

Contents

Foreword

Ken Baake has produced groundbreaking work in the rhetoric of science with his study of meaning construction at the Santa Fe Institute, (SFI) a think tank specializing in complexity theory, computer simulation of complex societies, and various approaches to the modeling of physical and biological systems. His work should be of interest to those who study the rhetoric of science and to those technical communicators who help scientists with their writing. There are other audiences who will find interest in his rich discussions of the practice of science. Philosophers or linguists who care about how language works through metaphor are another likely audience. Those who work in the more theoretical areas of scientific practice will benefit from thinking about the interaction of language, theory, and scientific work. Those rhetoricians interested in applied theory will find much to stimulate their thought, since the work is so carefully grounded in Classical and contemporary rhetoric, drawing strains from semantics to develop a theory of harmonics.

Baake is particularly interested in how metaphors come to have currency, what their status is for both insiders and outsiders to the community of scientists, and whether metaphors might more appropriately be considered stylistic or meaning constitutive. His approach combines ethnographic, participant-observer data gathering with close analysis of texts from the Institute. Interviews with a number of scientists and others associated with the Institute lend texture to his analysis, as do his various assumed personae at the Institute as a reporter, web author, and commenting observer. Baake's work demonstrates strong command over several literatures, including the language and rhetoric of science, metaphor as treated within philosophy, and the emerging disciplines of chaos theory and modeling of complex systems.

One of Baake's primary contributions comes from the light he sheds on the relationship of verbal to mathematical thought and expression. He nicely explores the preference for mathematical modeling, but demonstrates how his subject scientists recognize the limits of mathematics, both for codifying their own ideas and expressing those ideas to their fellow scientists and interested people from outside the Institute. He explores their deliberate and cautious regard for metaphor, and uncovers a range of opinion toward the use of metaphor. Baake develops a theory of *harmonics*, his own metaphor borrowed from the field of musical composition, to discuss the ways that metaphors gather meanings about themselves and become larger, more encompassing, and sometimes resonant with too many meanings to be useful as conceptual, theory-generating tools. He rightly characterizes metaphors as containing a range of meanings, with overtones creating dissonance that will be perceived as more or less tolerable, depending on the scientist's awareness, the positioning of that scientist with regard to contemporary competing theories, and appreciation of the range and robustness of the metaphor.

In a brilliant chapter on the term and construct *complexity*, Baake draws a tight focus on this single term and rehearses its life history across various discourses, showing how it gathers and loses its power to create and sustain insight. The chapter is enlivened by Baake's access to key thinkers on complexity, and by their own insights into the ways a special word takes on new meanings. The debate, here and throughout the book, centers on whether a term such as "complexity" can indeed be considered a metaphor, and if so, what distinguishes it from other, more formal metaphors, such as those with clearly distinguishable tenor and vehicle.

Baake's engaging presentation renders this work a treat to read. His years as a journalist are complemented by his work in business and economics. Baake is able to write in ways that engage and carry the reader on a highly enjoyable excursion. His prose is smooth and progressive; his arguments are well cast and a delight to follow. The text is populated with the voices and perspectives of his subjects, always presenting them as thoughtful and complex, even when they take diametrically opposed positions. Throughout the work, Baake shows respect for his subjects and a deep curiosity about their intellectual lives as well as the everyday work of science. When the science (or philosophy or rhetoric) gets difficult, Baake writes in ways that are especially inclusive, welcoming the reader into new territory with reassuring explanations and careful exposition.

Stephen A. Bernhardt
Andrew B. Kirkpatrick Chair in Writing and Professor of English
University of Delaware, June 5, 2002

Acknowledgments

Since the underlying theme of this book is that knowledge is developed socially, it should not be surprising that its author feels tremendous gratitude to many people who have helped shape the project. I have been working on this project since 1997 when it grew out of a science journalism and technical writing internship at the Santa Fe Institute. Five years later, with the help of scores of friends and colleagues, the project is now complete.

I offer thanks first to members of the Institute, who accepted yet another ethnographic study of their site and who took time to reflect deeply on the questions I was asking. Without this reflection, I would not have gained the insights presented here. I offer a special acknowledgment to Ginger Richardson, SFI program director, for her continued support during my association with the Institute. Her dry sense of humor, reflective nature, and ability to make things happen made it all possible.

This work evolved out of doctoral studies in the Rhetoric and Professional Communication program at New Mexico State University. As dissertation chair, Stephen Bernhardt shepherded the text through with thoughtful comments and enthusiasm. Steve set the dissertation on the course to become a book by nominating it for an award at the Conference on College Composition and Communication. I could not have completed the research without the firm grounding in rhetorical theory offered by David Fleming, Chris Burnham, and Stuart Brown who taught me that Plato's method of tackling philosophical challenges through problem-posing and focused dialogue still speaks to a postmodern world. Carl Herndl's thorough instruction on qualitative research gave me the confidence to take such an approach here. Tim Roth and Elba Brown-Collier at the University of Texas, El Paso opened my eyes to the ways that economists approach problems of knowledge.

As I began converting the project to a book, I received lots of support from colleagues and students in the English Department at Texas Tech University. Carolyn Rude made my entry into the ranks of assistant professor in Fall 2000 so smooth and non-threatening that I was able immediately to get down to this work. All new faculty members should be lucky enough to blossom amid such warmth and intelligence. Sam Dragga fine-tuned the all-important proposal letter to State University of New York Press. Many graduate students read different drafts of the work and tested it for usability in graduate classrooms. I offer thanks to Charlsye Smith for an early reading and vote of support, and to students of my Spring 2002 class in the Rhetoric of Science. I am particularly indebted to Kathy Northcut, who offered a page-by-page critique of the work in its later stages. She suggested many subtle, helpful changes.

Charlotte Kaempf at the Institute of Water Resources Management, Hydraulic and Rural Engineering at the University of Karlsruhe in Germany, gave the manuscript a close reading from a scientist's perspective and assured me that these problems of word meaning are by no means unique to the Santa Fe Institute, nor to the English language. In an email response she told me:

> "Whenever scientists are involved in or participate in interdisciplinary research projects they are bound to face language problems, partly due to disciplinary differences in meaning of terms, and partly due to use of metaphors for concepts and phenomena that they are about to analyze, simulate jointly. The situation at the Santa Fe Institute given here is representative for any science research group working on the 'frontier.' "

The folks at SUNY Press have been supportive and cheerful in a time when the academic publishing industry is under intense pressure. I particularly want to thank Ron Helfrich, Priscilla Ross and Judith Block at the Press, along with series editor Jim Zappen and the three anonymous reviewers. Their clear explanations for how to revise the project brought the best out in my writing. Bernadette Longo, whose book *Spurious Coin* preceded this one in the series, has been supportive throughout, restoring my confidence when it lagged.

I cannot offer enough thanks to my wife, Laura, and children, David and Rebecca, for their continued love and support during the time I have been writing. They made it possible for me to delve deeply into these intellectual questions; such contemplative time is a privilege not everyone is lucky enough to have. Finally, I thank my parents, Rick and Eileen, for passing on their senses of humor and love of music—two traits that have added immeasurable joy to my life.

1

Introduction: The Problem with "Rules"and Why Words Will Not Sit Still

A s a scholar of language, I knew I was onto something when I watched political scientist Elinor Ostrom speak before a group of scientists at the Santa Fe Institute in the summer of 1999. The scientists sat at a horseshoe-shaped table, eyes glancing back and forth from Ostrom to her slide presentation at the front of the room. Outside, cumulus clouds were visible as they puffed up into the deep blue sky over the valley below the Institute, a remodeled adobe ranch house. Scientists of various ages and areas of expertise sat alertly and listened as Ostrom explained her research with students at Indiana University. The students had been issued tokens worth money and asked to allocate those tokens among the group. Ostrom's research was grounded in game theoretical economics, which mimics real-world situations by requiring subjects in a group to make individual choices based upon perceived rewards. Ostrom concluded from her experiments that public policy issues—such as how to allocate ground water among farmers—should not be decided solely by a centralized governing body. Allowing the farmers to organize into smaller cooperatives to decide irrigation rights and rules is an effective alternative, she concluded, noting that, "Rules become the tools that humans use for self-organization."

Yet, when it came time for the question and answer portion of her talk, Ostrom found that much of the discussion did not focus directly on her recommendations, but was given over to a spirited debate about what the word "rules" meant to her model. For biologists, "rules" suggest regular observable behavior in an organism, Ostrom explained later. For example, a species of fish that always returns to the same place to spawn could be said to be following a rule. In the same way, I thought, perhaps even the clouds building outside were following a rule to form when the temperature, humidity, and barometric pressure reached certain levels. In contrast, however, to the meaning that biologists and other natural scientists assign the term, "rules" for political scientists are a description of action that is permitted by a social group. They require some kind of conscious choice among autonomous agents, and they are enforceable. The discussion became so mired in this semantic debate that one researcher in the audience was prompted to exclaim that for the purposes of discussion, "we just get rid of the word 'rules' because it is confusing the issue."

Of course his request, spoken in a moment of obvious frustration, was impossible to honor. Words and the problems of semantics will not go away even for scientists. One could ask whether useful discussion following Ostrom's presentation was sidetracked by the semantics debate, or whether the attention given to a single word led to a deeper understanding of the policy questions she was grappling with in her talk. But one could not ask for these semantic labyrinths to simply straighten out. Perhaps much of the best intellectual work at the Santa Fe Institute and other science centers comes out of the impromptu debates among researchers from different fields over the meanings of various words.

The story of Ostrom's talk exemplifies the constant fascination with language that I have observed at the Santa Fe Institute. The 18-year-old Institute is a think tank of scientists from various disciplines who are studying the relatively new science of complexity theory, which asks how order emerges in the world. To some language theorists, the word "rules" in Ostrom's talk would be functioning as metaphor, borrowing connotations from one field or discipline to lend meaning to another. An uncomplicated entry on metaphor from *Webster's New Collegiate Dictionary* defines it to be "a figure of speech in which a word or phrase literally denoting one kind of object or idea is used in place of another to suggest a likeness or analogy between them" (722). What confounded the scientists who were in discussion with Ostrom is that they could not agree on which fields should supply terminology definitions. They could not control the metaphoric process or the connotations that a simple word evoked among members of a mixed audience. Perhaps they thought that the word "rules" should have an exact literal meaning, but they could not get such a meaning to sit still long enough to ground a discussion.

Metaphor is the rhetorical issue that dominates the Santa Fe Institute. Erica Jen, the SFI's then vice president for academic affairs, stated in one interview with a Santa Fe magazine that, "Metaphors in science are powerful because even when you can't pin them down mathematically you relate to them on an emotional level; they get you moving" (Jarrett 16). I heard about problems with metaphor on the first day I walked into the Institute in the summer of 1997. As a scholar of rhetoric and technical writing, I was offering to help scientists write text that would summarize their research for the Institute's Web page. In the first conversation I had with Jen, and in most subsequent ones, discussion of metaphor came up. Institute members are on a constant quest that requires a delicate balancing act: they are looking for new metaphors that can stimulate productive scientific thinking, but without distorting that thinking. The SFI lexicon is filled with colorful terms like "spin glass," "complexity," "fitness," "compressibility," "emergent design," "autocatalytic sets," "sand pile catastrophes," and "cellular automata." These terms conjure up rich visual images and prompt recurring discussion among scientists over meaning. Some of these terms may be metaphors, while others may not. A useful goal for any writer who works with scientists is to help them to consider what qualities are present in metaphoric thinking. This goal recognizes that metaphor has both special powers and special dangers that must be understood before metaphor can be used wisely in directed scientific research. These SFI terms also have connotations outside the walls of the Institute that add to the Institute's mystique among the lay public, but also exaggerate the applications that may be possible from complexity science.

The purpose of my research is to examine language use at the Santa Fe Institute, specifically as it relates to rhetoric and the trope of metaphor. I am asking broadly what role rhetoric and metaphor plays in this kind of science and how scientists feel about language issues. While I have approached this project with many research questions, perhaps they could all be summed up with one: "What can the rhetorical challenges confronting members of the Santa Fe Institute tell us about the role of rhetoric—specifically metaphor—in science?" Embedded in that question are other questions, such as one I was asked by Ginger Richardson, the program director at the Institute and my closest contact there. She noted that problems of meaning confront all human beings and all discourse communities. "Is there something unique about the SFI that makes our language problems different?" she asked. I hope this project sheds light on that question and, in the process, helps those who attempt to forge scientific knowledge in interdisciplinary settings communicate better as they explore new theories.

I wrote this book with three primary audiences in mind. One comprises scientists, especially those involved in theory building, who find that the choice of the right words can have major implications for how a theory takes shape and disseminates throughout the scientific community. A second comprises technical writers who use their rhetorical skills when working with scientists to develop, edit, and deliver knowledge. My definition of a technical writer here is loose: for example, academic scholars who write about science from various disciplinary perspectives and science journalists are practicing variations on technical writing. A third audience comprises those with a keen interest in the philosophy, sociology, and history of science. I draw heavily from these disciplines for most of this analysis.

Both scientists and technical writers share a fundamental if unstated assumption, which is that reality can be transcribed by use of symbols and, hence, made more comprehensible and useful to human beings. The symbols can appear in mathematical formulae or in ordinary sentences; what is important in this assumption is the notion of transcribing reality, representing it in a way that is true to observation and is useful. A warrant that drives this assumption is that the scientific method must close in on reality, must define it ever more precisely to remove unintended implications or spurious associations. If the word "rules" were emitting noisy associations then it would be the job of Institute scientists and writers to fix the coordinates of its meaning more precisely. Yet, as Bernadette Longo makes clear in *Spurious Coin*, her book about science and technical writing, transcriptions of science as reality always emit spurious signals. Longo's research suggests that scientists see such spurious signals as inevitable but also lamentable. Perhaps Santa Fe Institute scientists realized that any discussion of theory that involved an evocative word like "rules" was destined to break down over confusion about which signal to tune in.

My presupposition in 1997 as I began a relationship with the Santa Fe Institute was compatible with implications from Longo's book: any understanding of reality that emerges from a debate over word meaning distracts the scientist from demarcating the precise and useful aspects of that reality. I assumed that these Santa Fe Institute scientists inadvertently stumble into debates over semantics, but prefer to find ways to avoid such debates in order to get the real business of science done. If so, then these scientists would use rhetoric and its figurative language begrudgingly, primarily to communicate with non-scientists. It would follow that scientists see language not as a tool for producing knowledge, but merely as a vehicle for communicating knowledge that has been adduced by other more empirical and rigorous means. Perhaps the near obsession with metaphor that I saw from the beginning at the Institute was temporary; once the scientists got beyond the limitations of language, then the real science could begin. Even as I reveled intellectually in debates, such as the one over "rules," I still believed that perhaps language could be made to sit still.

I was influenced by the biases of post-positivist philosophy as it is commonly understood today, which holds that science is the testing of hypotheses by formal empirical methods. Language theorist Robert Hoffman, citing various researchers looking at metaphor in science, points out that the strict empirical perspective would see metaphor at best as a heuristic, to be filled in later with a real theory. "At worst it is an irreal fungus doomed eternally to a prescientific twilight zone," Hoffman adds, in summarizing this dismissive view of metaphor. "Any good theory should be literal and precise" (394).

Deep down technical writers working among subject matter experts often feel a slight inferiority complex. I too felt it upon entering the Santa Fe Institute, believing that most scientists still adhere to Francis Bacon's argument: if you look at the world long enough, the answers will emerge without the midwifery of language. A glance at scientific journals reveal page after page of mathematical symbols building to what appears to the layperson as an incontrovertible proof of some theory of reality. Having come out of master's program in economics, where the term "mathematical rigor" was held in God-like veneration, I assumed that most scientists believe reality to be knowable only through excruciating mathematical analysis. Vague language—my main area of expertise—offered at best a shadowy image of the mathematical truth.

Although language may be inherently vague, I wondered if it could not be made more precise by careful attention to definition at the outset of scientific research. As a technical writer, I felt charged with the goal of helping Institute scientists use language—even metaphoric language—in an exacting way that would transcribe reality by suppressing spurious meanings, thereby allowing precise ones to ring clearly. It seemed logical that if Ostrom had first defined her terms, or even used subscripts to distinguish one type of rule from another, certainly she would not have lost time to a semantics debate. A sensible goal for a technical writer at a place like the Santa Fe Institute would be to help its scientists to use language more rigorously, if not elevating it to the status of mathematics, at least making it a worthy junior partner.

Throughout this book, I reconsider these presuppositions, gradually revising the hypotheses that have emerged from my research. After working with those Santa Fe Institute scientists and interviewing them about language issues, I have come to realize that these scientists value metaphor and other rhetorical devices even as they are uneasy using them. They depend on metaphoric language to generate theory across disciplines and, equally important, to make their studies seem exciting, cutting edge, and worthy of publication and funding. At the same time, scientists try to distance themselves from metaphoric expressions when they want to appear rigorous and far removed from the social fray that discursive language inspires. Hence, these scientists rely on metaphor, while at the same time trying to rise above it. Their paradoxical response is similar to that of the Greek philosophers, whom we will encounter throughout this book. Recall that these philosophers coveted rhetorical skills for their inherent power to produce knowledge and to persuade, and yet they attempted to keep a safe distance from rhetoric so as to not appear lustful for the fruits of such eloquence.

The history of science is full of debates over metaphors, such as debates in ninteenth century physics over what was meant by a "field," whether energy and heat were actual entities or merely evidence of molecular motion, and whether electricity was a kind of fluid. Language, particularly metaphoric language, resonates with meaning and implication. A scientist may use a word to mean one thing, but she cannot prevent others from hearing different signals from the word—even if those signals cause the hearers to question the essence of the speaker's argument. I hope that the reader learns from this project, as I have learned, that such debate over word meaning constitutes rather than impedes scientific knowledge. This is a question of epistemology, which is a fundamental problem of philosophy that has occupied thinkers for several thousand years. I argue that metaphor is a tool of epistemology and that technical writers trained in rhetorical theory can help scientists to manage that tool.

When we read about knowledge we often find several key concepts in the proximity; these include observation, belief, reason, and reality. Knowledge for my purposes is reasons and beliefs that link an observer to reality in a consistent way. That is, what we know today we should also know tomorrow, even if tomorrow's knowledge is not synonymous with today's, but only aware of it. Plato's dialogue *Theaetetus*, the foundational text in epistemology, suggests that knowledge is the shaping of belief, derived from perception, by reason. Plato uses a metaphor of knowledge as clay shaped by the human hand into bricks (147a–c6). Whether knowledge as clay exists prior to that shaping is unresolved, but the key point for this study is that scientific knowledge is the result of some kind of scientific reasoning process—a scientific method—that is applied to observation. Of course, as philosopher of science Helen Longino points out, direct experience is not a prerequisite of knowledge; most statements about reality are accepted even if not directly experienced by the knower (148–156). (I know that World War II happened even though I wasn't alive then. Scientists know that unstable uranium atoms can be driven to a chain reaction even if they have never witnessed nuclear fission.) Thus, we accept the shaping of clay that we have not seen happen. Since I will draw parallels to music theory, this Platonic image of brick making may be confusing. An image consistent with my argument would be that of scientific knowledge as musical notes assembled into some kind of meaningful and evocative pattern.

HARMONICS: USING MUSIC THEORY TO EXPLAIN HOW METAPHOR WORKS IN SCIENCE

A scientist using a metaphor like the word "rules" may intend for audience members to hear the fundamental meaning of that word in her field, "a codified, consensual social structure." She cannot prevent others from hearing different meanings, however, such as "a regular pattern of involuntary biological behavior." Borrowing from the physics behind music theory, I use the term "harmonics" to refer to the various implications or signals that a word carries. Musical harmonics are tones that sound in addition to the fundamental tone when a note is played. So, for example, when a violinist plays a note at middle C, nearly twenty different tones may sound simultaneously (Seashore 98). These tones, known as "partials" or "overtones," fall at various frequencies above the main tone, meaning that each note is a blend of musical intervals. It is the unique combination of overtones associated with each instrument that give it a subjective quality known as "timbre." The different array of overtones is what distinguishes a clarinet from an oboe, for example.

Most instruments offer this collage of tones every time one note is played, although the ear usually does not distinguish among the various overtones. Intuitively, we can assume that a metaphor functioning as a musical note would convey a fundamental tone, denoting the primary meaning, but also other tones at regular intervals, the harmonics. The implication of this way of explaining metaphor is that some meanings resonating from a word will sound consistent with the listener's expectations of sounds, while others will not. A "rule" as a sociological phenomenon may sound appropriate to the social scientist, but the harmonic meaning suggesting a biological aspect may be disconcerting to that same scientist. The social scientist in this example must either find a way to suppress the unpleasant sounds, adapt her sense of sonority to accommodate the new sounds, or reject the fundamental tone altogether. In this way, metaphor (as a tone on an instrument) changes the meaning of the whole piece of music; it "constitutes" the piece. This, I argue, is how metaphor constitutes theory in science.

The analogy of words to musical notes that we have seen so far implies that metaphors stand alone as individual notes filled with multiple meanings. Yet, language and music are about relationships, intervals between word meanings or tones. Although one word or one tone does carry unique sounds or meanings, the real focus in my theory of metaphor is on the way that words interact in building and refining scientific theory. In music, when a tone is heard among others or immediately after others, our ears make a judgment about the overall sonority. The fundamental sound of individual bells sounding a close succession of notes can be disturbing, for example, because so many dissonant harmonic overtones can make it difficult to distinguish the fundamental pitch ("Harmonics," Sadie). Likewise, pianists rarely play bass notes together because these notes generate many harmonics that together sound discordant.

In examining the way scientists use metaphors, I have found that metaphors tend to be accepted if they sound sonorous amid the overall paradigm in which those scientists operate. So a metaphor and all of its meanings must fit into the theory in the same way a note and all of its harmonics must fit into a melodic phrase. A scientist who is developing a theory about social activity, or any activity that involves choice among the agents, will hear that the word "rules" sounds consonant with the theory. Dissonance appears when the word is pushed to apply to situations where behavior is involuntary. (How can something follow a rule if it has no choice?) The dissonance need not destroy the theory; instead it may cause scientists hearing it to accept new implications of a word within that theory just as Beethoven's genius pushed audiences of his day to accept sounds previously thought to be jarring.

The musical way of understanding metaphor seems appropriate because it is consistent with knowledge that language originated as an oral/aural system of communication. Obviously my argument is not trying to carry the analogy with music so far as to suggest that word meaning depends on the actual sound of a word. This is not a theory of *onomatopoeia*, where a word literally sounds like what it means. I am arguing that any word conveys multiple meanings, whether we hear those words spoken, we read them, or simply think of them in our minds. These meanings consciously or unconsciously affect the way we accommodate that word into a larger syntactical cluster—the sentence that attempts to represent reality. Metaphors by definition are filled with associations and, hence, they add many dimensions of meaning to any sentence in which they appear. These dimensions of meaning are what I liken to musical harmonics.

The physics of harmonics has occupied philosophers at least as far back as the Classical Greeks. Parts of their analyses of harmonics are relevant to my project because they explore the ways in which musical and philosophical meaning is a combination of sensory perception and reason. Spanning the more than 800-year-period from the Pythagoreans and Plato (circa 400 B.C.E.) through Claudius Ptolemy (circa 140 C.E.) to the later theorists, Greek philosophers held forth on whether music had its effect primarily because of the inexplicable delights of sound that it brought to the ear or because of the rational, mathematical relationships inherent in the frequencies that conveyed those sounds. The debate in science over whether knowledge is what we induce empirically or whether it is what we reason deductively has it origins in the Greek analysis of the effects of music. Extending the debate to metaphors functioning as musical notes, we can see that our response to a metaphor in science can emerge from our rational analysis of word meaning, but also from how words strike the ear in combination with other words. Language originated as a spoken system of communication, so even when we read words we hear them in our minds. Meaning is reasoned but also sensed aurally.

In Plato's dialogue *Timaeus* we find a discussion of the crafts-man of the world—a divine mind—dividing the soul of the world into harmonious intervals, which have a mathematical basis (36a,b). These intervals are modeled after the perfect "Forms," from which the elements of reality are modeled. In the *Cratylus*, Plato describes words as another means of dividing the world in the same way that a weaver's shuttle is used to divide yarn. "So just as a shuttle is a tool for dividing warp and woof, a name is a tool for giving instruction, that is to say for dividing being" (388b13). In this dialogue, the craftsman god uses names to divide reality in the "natural way." For Plato the process of naming the world would have to follow the nat-ural harmonic intervals into which it was divided. Metaphors would be measured by how consonant they sounded in the context of such a rational system.

This consonance could be determined by spirited debate in the same way we saw Santa Fe Institute scientists work out whether the term "rules" was acceptable for the theory they were discussing. Debate over word meaning was the prime method of clarifying knowledge for the Greek philosophers. Plato in the *Cratylus* argues that correct words are divinely inspired. Yet, such words are refined by those who ask questions—the dialecticians (390c,d). Plato's suspicion of rhetoric would have prevented him from trusting this semi-divine task to rhetoricians alone, but the point is nonetheless prescient. Hence, we can conclude that knowledge making in science is a cogni-tive process by which scientists as Platonists attempt to find the appropriate theories, or names, for reality. Yet, this is a far more social process than perhaps Plato would have acknowledged; scientists con-tinually sound out their theories against the background of knowledge and beliefs that have been accepted as sonorous by their peers.

Scientists and those who help write science may have trouble accepting that knowledge is in any way a function of what sounds good. Indeed, this argument would have troubled Plato. He argued that the effect of harmonics on the ear was a less accurate indicator of consonance than was the mathematical relationship among the frequencies of the tones. While sympathetic to the early Pythagoreans, Plato criticized those who relied on the ear and who "don't investigate, for example, which numbers are consonant and which aren't or what the explanation is of each" (*Republic* 531c). Andrew Barker, a professor of Classics, shows that for Plato and for anti-empiricists ever since, the senses were too easily deceived to be reliable. "The distinction between the beautiful and the merely pleasant is at least as old as Plato, as is the tendency to associate beauty with rationally intelligible form, pleasure with mere titillation of body parts" (265). This Platonic criticism of the senses wends its way throughout the history of scientific knowledge. Traces of it are evident in the interviews I present with the Santa Fe Institute scientists. We will see that these scientists rely on metaphor, but at the same time view it as less trustworthy than mathematics, albeit potentially more titillating.

Aristoxenus, a follower of Aristotle who lived in the late fourth century B.C.E., argued that such devotion to reason over the senses was illogical. His philosophical investigations suggest that no intrinsic mathematical law can determine which harmonics sound pleasing any more than the description of an animal can serve as a substitute for the actual animal. Aristoxenus and his followers pointed out musical intervals that sounded good to Greek ears (and would still to ours), but which were rejected by Pythagoreans because they could not be expressed in appropriate mathematical ratios (Mathiesen 379).

Claudius Ptolemy, working at the Library at Alexandria, developed an extensive theory of musical harmonics that accepted arguments from the followers of the Pythagorean School and Plato about the rationality of music and, also, arguments from the more empirical side. As Barker reveals, sound for Ptolemy is *pathos* of the air, a phenomenon not ruled by reason. Yet, Ptolemy recognizes that the senses by themselves are unreliable because their perceptions are not consistent from person to person, or even within the same person over time. Hence, Ptolemy concludes that the science of harmonics requires both reason and perception. Barker offers this summary:

> The senses, then, are insufficiently exact instruments to make the precise discriminations required by the scientist. Reason, on the other hand, is powerless on its own. It has no independent

access to the data, but must take from other sources the rough and ready information they give about the contents of the world. (19)

It would follow that unreliable data are inevitable in science and, likewise, that potentially discordant harmonics are inevitable when the scientist uses words to interpret, represent, and organize that data.

Musical harmonics for Ptolemy were not merely sounds, however, but the relations among sounds. "No one thing, be it a note or a distance, can be concordant in its own right, but only in relation to something else" (Barker 93). This Ptolemaic dictum is central to my argument that metaphor functions in the context of many words, which together make a claim about reality. Knowledge emerges through the perception of a signal—metaphor, in this case—that evokes associations/harmonics. We accommodate those associations by listening to and reasoning about them in relation to the other words in the discourse cluster. A metaphor in a biological context conjures up different associations than it would in a political science context, almost as if the metaphor were "played" on two different instruments.

Discussion over associations builds new theory at places like the Santa Fe Institute. Revisiting the age-old debate about whether metaphor shapes theory or merely decorates it, I will show that it constitutes theory by ringing forth with various signals, various meanings. These force scientists to confront the implications of any theory. Paradoxically, imprecise language sometimes leads to a greater and more thorough understanding, which is, after all, the goal of interdisciplinary research. Despite the occasional moment of frustration over semantics, interdisciplinary scientists need their terminology to be vague enough to embrace a variety of meanings, the variety of guises that reality may wear. A technical writer who works with scientists must come to realize that despite the writer's strong instincts to the contrary, the goal of scientific communication may not always be precise representation of subject matter. The writer's job is challenging in such a setting because he must intuit from the subject matter experts how much vagueness is acceptable, even desirable, in a text. The goal in writing is to choose words that, while being accurate, also ring forth with enough degrees of freedom to be coherent with reality's complex nature.

Rhetorician Jeanne Fahnestock makes precisely this point in her analysis of the debate in archaeology about when the first humans entered the New World. Rather than debating to resolve the argument, the goal may be to continue the debate. As Fahnestock writes, "The true exigence may favor not agreement but continuation of the controversy" ("Arguing . . . " 65). In this context, the writer's contribution to science is not in suppressing metaphor harmonics, if that were even possible, but in helping scientists recognize how those harmonics resonate within what might be called the "tonality of meaning."

STRUCTURE OF THE BOOK

This project begins with analysis of my own rhetorical challenges as an occasional freelance technical writer for the Institute. The main part of the research then involves interviews with many scientists affiliated with the SFI. In these interviews the scientists talk often about metaphors in their field that are evocative, but often frustrating because of their tendency to oscillate among different meanings. I also draw freely upon texts produced by those scientists for examples of rhetorical strategies, and to explore how those scientists deal with problems of language meaning and audience in their own writing. I have tracked the use of a key term at the Institute, the word "complexity," examining how it has moved in and out of science and popular discourse over the years. In my research I have repeatedly engaged questions put to me by Institute members. Some of these questions go to the heart of metaphor studies, such as, "What is a metaphor?" and "What is a metaphor's purpose?" One version of these questions came up, for example, when I was referring to "complexity" as a metaphor and two people at the Institute in separate discussions asked why it would be a metaphor, as opposed to "just being a word." I could not answer the question. Nor could I say then with conviction why it matters that scientists recognize metaphors when they use them, or why they should accept the power of rhetoric. Yet, somehow I have always believed it does matter. That belief has motivated my work.

Researching these core questions has forced me to take a position between two extreme poles of thought. One, identified with I.A. Richards, and perhaps with George Lakoff and Mark Johnson, assumes that all language is metaphorical, even when not intentionally so. Another, drawn from Aristotle's founding work on rhetoric, assigns a narrower role to metaphor as the intentional use of language by gifted or skilled rhetors to highlight resemblances across fields—an action undertaken for specific epistemological or persuasive purposes.

At one extreme, metaphor is pervasive, inescapable, and involuntary. It is the essence of word formation within language. At the other, it is sporadic, ornamental, and deliberate—something layered onto the surface of language. The first part of this book attempts to close in on an understanding of metaphor as an epistemological concept by exploring these theories in the larger context of rhetoric and science.

Chapter 1 continues with a brief description and history of the research site. The purpose of this chapter is twofold: to situate the reader at the Santa Fe Institute and to outline the research that I have conducted there. Chapter 2 is a narrative essay that reviews the writing projects I had been involved in at the Institute prior to beginning the interviews. Technical writers reading this book will no doubt recognize some of the joys and frustrations that come when trying to produce various texts for a group of brilliant but at times impatient subject matter experts. Here I explain my informal participant-observer role as an occasional writer, and offer insights into the rhetorical challenges I have observed while researching and writing these projects. These observations came during visits to the Institute to attend meetings and gather information for the writing projects, and during email and telephone conversations with Institute members throughout the research, writing, and editing stages. The observations I made while writing for the Institute led to the questions that launched the formal study.

Chapter 3 plumbs the depths of the literature related to metaphor theory and the rhetoric of science, examining the relationship between questions from those two fields. I situate the reader in the ongoing debate over the roles that metaphor and other rhetorical devices play in the development and dissemination of scientific knowledge. From here it becomes possible to explore these language issues at the Institute.

Chapter 4 pulls together and analyzes the findings from the formal study. It involved interviews with seventeen people affiliated with the SFI, including resident scientists, visiting scientists, and administrative personnel. In this chapter I explore the function of metaphor in theory building as seen by the SFI people. Because the scientists frequently considered metaphor (and all discursive, lyrical language) in contrast to mathematical expressions of knowledge, I found it worthwhile to do the same. Hence, the chapter also explores a fundamental debate in the philosophy of science over the indispensability of mathematics to scientific knowledge. Scientists in general are sympathetic to the Platonic argument that truth is objective and that it can be accessed only by objective epistemologies, of which mathematics would be supreme. Anyone who writes in a scientific setting will find this discussion of the tension between mathematical knowledge and lyrical knowledge to be helpful, for this tension lies at the center of scientific practice.

Hence, as the chapter builds from the semantics debates ever present at the Institute, such as that over the word "rules," it asks whether replacing such language with perhaps the more pure representations of mathematics could filter out the unwanted harmonics of metaphor. I show that mathematical and metaphoric thought processes are often intertwined and impossible to isolate. Scientific intuition cannot be reduced to one epistemological method. Here is where my image of word harmonics among scientists takes full shape and leads to the claim that such harmonics are not spurious, but theory constituting. I dip into the theoretical literature again and draw on the theories of additional philosophers, rhetoricians, and scientists when necessary to help make sense of the issues that emerged in the interviews.

Chapter 5 develops a secondary theme that emerged from the interviews—the challenge that scientists face in presenting their findings to other scientists and to the outside world. This involves issues of style and eloquence and the best way to communicate with scholars in different fields; rhetoricians know this as the problem of "incommensurability." During the interviews, several scientists expressed frustration that communication was difficult even at a science think tank whose mission was interdisciplinary research. The problems of metaphor harmonics can make it difficult for one scientist to know what another means. Yet, not all problems of communicating across disciplines are so abstract; these interviews reveal that Institute scientists often face inertia among their peers when trying to develop interdisciplinary research. The problem often is as simple as trying to find a journal that will accept an article hewn from disparate fields, or trying to find grant agencies that will fund such research. Delivering ideas can be especially confounding for scientists when they try to rewrite their work for a general audience, as my analysis of the work of Institute biologist Stuart Kauffman reveals. The technical writer who works with such scientists should take away some insights into how to help them to navigate the rhetorical challenges involved in delivering knowledge to different audiences.

The challenge of communicating across disciplines is not new to science; such challenges appeared in the eithteenth and nineteenth centuries when science as natural philosophy branched out into various fields. These fields developed specific focuses on the natural world. Practitioners in each field defined their activities according to their focus, whether it was at the atomic level, the molecular level, or the organic level. The names "biology," "physics," and the like circumscribed an area of focus for each field. At the Santa Fe Institute, scientists are attempting define a new field called "complexity," which crosses many fields. Chapter 6 attempts to determine the scope of this field by taking a close look at the etymology of the word "complexity." This history shows how meaning can change in science, and how a word like "complexity" can accompany and perhaps prompt dramatic shifts in scientific thinking. Here I get closer to resolving the questions of whether "complexity" is a metaphor and why it has been such a powerful word. The essence of this chapter is an exploration of "complexity" as found in The *Oxford English Dictionary*, where several pages are devoted to description of the term in its various forms and examples of how it has been used since Classical times. I also sample a few of the more than 200 texts that refer to "complexity," as listed in a university's electronic catalogue. The chapter also includes a few interview comments specifically related to this key term. As this chapter shows, a writer working with subject matter experts often can illuminate knowledge for those experts by revealing the histories of terms they hold dear.

Chapter 7 wraps up the book by revisiting the central questions that have been running throughout. I conclude that SFI rhetorical challenges in some ways may be more acute than the challenges facing more traditional scientists, but that those challenges are not fundamentally different. I claim that those rhetorical challenges are not impediments to good science, but instead are essential catalysts for novel thinking. Here I reconsider the ways in which metaphor adds value to science. The chapter offers general recommendations for how technical writers can assume their place in such science organizations. The underlying warrant is that technical writers trained in rhetorical theory belong in centers of science; the subtle understanding that technical writers possess of how words and their harmonics work can help scientists develop and fine tune their theories through the process of writing.

I must acknowledge at the outset that the conclusions reached through this research on metaphor and other rhetorical challenges in complexity science certainly cannot stand the test of doubt that defines the post-positivist system as it is understood today. I have made claims that cannot be falsified, for how could anyone find a case where language is not necessary to shape knowledge? That person would have to make his case without using language, which is impossible. Still, the qualitative research here is credible and honest. I have attempted a style of research known among ethnographers as "naturalistic inquiry," which aspires to standards of trustworthiness. As codified by ethnographers Yvonna Lincoln and Egon Guba, naturalistic inquiry seeks to depict a culture by allowing insights to emerge over time from among its members. A researcher should listen to the members when designing a project and should continue listening throughout the research and writing stages, not just during the so-called "data collection" stage. In doing so the researcher pursues important questions toward meaningful insights and continually modify his hypotheses. The researcher and members are partners; the resulting text becomes a kind of Platonic dialectic, from which insights emerge.

This concept of "emergent design" (Lincoln and Guba 225) is similar to the idea of emerging order found at the Santa Fe Institute. My project has evolved and taken different forms over the few years that I have been involved with the SFI. It is very much an emerging project. As I have written material for the Institute, beginning with my "Initial Impressions" report on the rhetorical challenges, I have made those texts available in draft stages to members. Responses from those members, including edited changes, have guided me in developing questions related to rhetoric at the Institute. Listening to members and reading works by various members made it clear that metaphor is not just my interest, it is a strong Institute interest. At one point late in my research, a rhetorical theorist asked why I had chosen metaphor in science as a topic of study. So much work already had been done in this area, she said, and other rhetorical devices are more interesting. I replied that because metaphor is what was on the minds of these scientists, it also was on my mind. Metaphor had emerged as an important question for these scientists, even though it may be passé for some rhetorical theorists.

Lincoln and Guba show that a naturalistic inquiry should aspire to standards of "trustworthiness" rather than to conventional positivist standards of "objectivity" and "validity." A naturalistic inquirer is not attempting to follow a strict method in order to validate a hypothesis that then can be accepted as approaching truth. A naturalistic project, instead, attempts to test findings through careful and open research to reveal insights that seem honest, believable, important, and not aberrant. A trustworthy project primarily must be "credible." This requires;

1. prolonged engagement, persistent observation, and triangulation

2. an external check on the inquiry process

3. refinement of hypothesis as evidence comes in

4. checking of new findings against archival data

5. the opportunity to check findings with the members who generated them (301).

From 1997 to 2000 I focused on the Santa Fe Institute, its research symposia, its texts, its key terms, and the spoken observations of many of its key researchers and administrators. Throughout the research I subjected my results to the credibility test by making them available to SFI members. "Member checking" is a term used in ethnography when the researcher gives the interviewees a chance to read interview transcripts and the draft analysis before publication, and a chance to modify their comments. The goal of member checking is to treat participants as people involved in the knowledge creating that goes on in research, not as objects to be mined for information. My most successful effort at member checking was in writing my final piece for the *SFI Bulletin* midway through my formal research. This column, published in 1999, served as kind of an abstract or executive summary of what I had discovered in my initial interviews: the Institute is interdisciplinary, highly philosophical, and abstract, but also burdened with great expectations from a public looking for big answers from science.

WHAT IS THE SANTA FE INSTITUTE?

Founders of the Santa Fe Institute set it up in 1984 as an independent, nonprofit research center dedicated to an interdisciplinary approach to science. The stated goal at the Institute was to break down the traditional academic barriers that often keep scientists of different backgrounds from working together. Hence, the Institute draws researchers from the natural sciences and the social sciences, including physics, biology, psychology, mathematics, economics, immunology, linguistics, history, and other fields. Arguably, this mix of researchers from various disciplines creates a greater need for metaphor than might be found at research centers drawing from a single discipline. The SFI brings these researchers together temporarily—in "a floating crap game," as Institute people are fond of saying—to nurture their ideas and techniques, and to study related patterns that appear across all these disciplines. Over the course of a year, the Institute hosts some 150 scientists for varying stays, with about 35 in residence at any time. A few are on multi-year appointments, a few are graduate students, and others are scientific visitors, predominately from universities in the United States and Europe.

Complexity science is the underlying premise that unites these disparate researchers. A fixed definition of "complexity" is elusive even for Institute members, but in general the theory suggests that chemical compounds, organic cell structures, Renaissance financial banking systems, as well as other physical, natural, and social phenomena all derive from an underlying order that propels them toward organization and adaptation. Therefore, researchers are constantly on the lookout for patterns, regularities, and structures behind all sorts of real world phenomena. Scientists often borrow the genotype-phenotype paradigm from biology to envision how underlying patterns become manifest at the surface level. In this biological way of viewing the world, the "genotype" is the code in the human gene that describes how the "phenotype," or individual body, is constructed. Yet, the question of how a code or set of rules is manifest in the evolution of all types of self-organizing systems is unresolved and subject to intense theoretical contemplation and debate. Even the relationship between the genotype and phenotype remains a subject of study.

Researchers collaborate at the Institute on projects dealing with everything from the communication patterns of ants, to the way information spreads across economic markets, to how the first replicating life forms may have emerged on earth. Institute scientists say that their exploration of these phenomena is helping define new research directions at major universities. This research in complexity science has been categorized by philosophers Steven Best and Douglas Kellner as one form of postmodern science. Best and Kellner argue throughout their 1997 book that postmodern science rejects the deterministic world view of modern science for a version of reality that is indeterminate, probabilistic, relative, chaotic, complex, and in a constant struggle between entropy and emerging order.

The interdisciplinary and cross-cultural ethos of the Santa Fe Institute is easy to discern from a visit to the Institute's center, perched on a hillside at a restored ranch house outside Santa Fe. For example, a visiting economist joining other scientists for lunch on the patio might share ideas with an architect who is seeking Institute help to study complexity in the design of buildings. She might hear discussions about molecular biology in French, German, Norwegian, and various languages besides English. Later, she might attend a talk about complexity in the stock market or about game theoretical approaches to policy making, which was the subject of the talk I referred to earlier. Our visiting economist might retreat to a computer workstation with graduate students to explore the evolution of an artificial life form they are modeling. During a late afternoon tea break, she might reconnect in the kitchen with fellow scientists to share information on her impressions. Later, back at her home institution, she could continue to share ideas with fellow Institute scientists by electronic mail and telephone, or return to the Institute periodically for seminars on topics such as the mathematical power-law relationship in earthquake incidence.

The visitor who arrives at the Santa Fe Institute expecting to see traditional experiments conducted amid laboratory beakers, microscopes, Van de Graaff electrical generators, and similar accoutrements of empirical science would be surprised. Most of the research at the SFI is on computer workstations, where artificial societies of agents and organisms are simulated. Stanford anthropologist Stefan Helmreich in his 1998 ethnographic study of the SFI notes that these computer-simulated worlds are known variously as "artificial life" or "cellular automata" or "microworlds." Scientists establish the agents present in the artificial worlds and the parameters for how those agents will interact among each other and the environment; the scientists then allow the computer to proceed through multiple iterations to see what happens to those worlds. These kinds of simulation systems have been used in science for decades, and are even available in popular computer games.

An example of one artificial life simulating system is the "Swarm" software, developed by former SFI researcher Chris Langton. In late 1997, Langton showed me one "world" he had simulated with Swarm—that of a 1,000-year-old community of Native American Anasazi people who had lived in the high deserts of what is now

Southern Colorado. To give the reader here some sense of what goes on in this elusive and difficult to envision science, I offer now a few paragraphs from an article I wrote for the *SFI Bulletin* about Langton's project:

> (T)hose researchers who have become comfortable with Swarm have found it invaluable. Tim Kohler, chairman of the department of anthropology at Washington State University, says Swarm has fully lived up to his expectations. The simulation software has led to intriguing insights into the ways in which early Native American peoples coped with changes in their environments. Swarm has corroborated theories suggesting that Anasazi cultures developed maize trading among households to contend with variable patterns of rainfall.
>
> In the Swarm Artificial Anasazi model, the terrain of Southern Colorado 1,000 years ago shows up on the computer as a green topographical map, crisscrossed with streambeds. Families show up as pixels of light. The model simulates life among the people using some thirty variables—birth rates, death rates, local topography, corn storage potential, rainfall, and the like. The computer races through each year in a matter of about ten seconds, allowing researchers to tweak one or two variables and rerun the model to see what changes occur. Researchers then study the survival rates among the pixels representing families.
>
> Kohler and co-researcher Carla Van West have found that families tended to be risk averse. They would share maize with other families in exchange for a promise to return the favor . . .. But when the entire region suffered from low maize productivity, families tended to hoard.
>
> Research on the Artificial Anasazi project suggests that environmental degradation forced the people to abandon their homes in the thirteenth century. But the Swarm model also suggests that one third of the population could have survived if they had redistributed themselves on the land. (Baake, "Swarm" 22)

In writing this piece I deliberately tried to explain how the computer simulates a living environment; I believed it necessary in the modern scientific tradition to emphasize method. Yet, much of the writing about SFI research, even for a general audience, occludes method or refers in vague terms to an "integrated approach" or "model building" or "examining" how adaptive agents behave. It is often difficult from reading Institute literature to know much about what a scientist is doing in a project. Helmreich's study suggests that many artificial life researchers have become so accustomed to the idea that the worlds they create on their computers are somehow real that perhaps they overlook the fact that they are engaged in a research method that involves setting parameters and observing simulated behavior. Perhaps, for the community of SFI scientists, writing extensively about simulation methods would be stating the obvious. Yet, for the outside reader, lack of attention to method makes it difficult to discern the boundaries between a scientist's thoughts and intuitions and his or her research results. The dichotomy between deductive, rational science and inductive, empirical science that originated in Aristotle and solidified in Bacon, Hume, Locke, and other Enlightenment philosophers is blurred at the Santa Fe Institute.

To make possible all of this SFI research, the Institute employs a support group of some 25 administrators and scientists who handle budgeting, publications, and related duties. Much of the funding for the Institute comes from National Science Foundation grants and from similar sources. Resident researchers typically stay for several months at a time. In addition, the Institute sponsors various summer programs that draw undergraduate students from all over the world. Each year, the Institute hosts summer graduate and undergraduate interns, sponsors a summer school problem in complexity science, and runs a two week "Complexity in Economics Program" for graduate students in economics. The Institute also helps to publish books, magazines, and articles related to current research, and several scientists have written popular science books based on their insights gleaned from SFI research.

The Santa Fe Institute has been the subject of many articles and books written by outside researchers and popular science and business writers who seem to be attracted to the eclectic and innovative research, but who at the same time also seem to be a bit suspicious. Several of these books are on display in a glass case near the entrance to the campus. In closing this section on the history of the SFI, it is worth mentioning briefly a few of the themes that have appeared in several of the popular science books. My purpose is to reveal the almost priestly aura that surrounds these scientists and their meeting ground in Santa Fe. I will return to some of these books later to explore their insights into metaphor at the Institute. Yet, for now the reader need only glean from the brief summaries the sense of profound significance that these authors ascribe to the Institute and its work, even when those authors are critical of that work.

The first popular book that was devoted almost exclusively to the Institute's history was M. Mitchell Waldrop's *Complexity: The Emerging Science at the Edge of Order and Chaos*, published in 1992. Waldrop, a particle physicist and writer for *Science* magazine, offers a narrative account of how various "old Turk" scientists, as he describes them, established the Institute in the mid 1980s. These veterans included various scientists from Los Alamos National Laboratories, including the former head of research, George Cowan, and several Nobel Laureates from various prestigious institutions, including physicists Murray Gell-Mann and Phil Anderson, and economist Kenneth Arrow. Waldrop is generally enthusiastic—at times even effusively reverent—about the interdisciplinary approach that launched the Institute. His book enjoys a mixed reputation at the Institute now, however; some see it as an accurate account of the Institute's founding, but others see it as an overly "gushy" work that focuses excessively on personalities.

The book by Stefan Helmreich I referred to earlier is called *Silicon Second Nature: Culturing Artificial Life in a Digital World*. This 1998 book is the product of Helmreich's doctoral research in anthropology at Stanford University. It considers the SFI as a site for cultural studies ethnographic research and offers many insights about the new "tribe" of artificial life scientists. Helmreich disdainfully paints a picture of the SFI as a male-dominated organization that is largely insulated from the outside world of Santa Fe, which includes a blend of old Hispanic cultures, new age artists, and gay emigrants. Helmreich explores language issues, writing that SFI scientists and others doing similar research have transformed the image of the computer from that of a tool of "bureaucratic rationality" to an organic metaphor for natural life (13). Helmreich writes that many of these scientists play the role of a paternalistic God in creating their "worlds" and in using religious terms to discuss their simulations. Some of the SFI people I spoke to about Helmreich's book acknowledge that he made interesting and often brilliant observations, but some say that many of his insights into the SFI culture are now dated.

Another book that considers parallels between SFI science and religion is George Johnson's *Fire in the Mind*, published in 1995, which narrates the author's ponderings and insights into the Santa Fe Institute and the current state of theoretical science in context of Northern New Mexico's mix of Catholic and Native American spiritual traditions. The book examines how information—whether it is the position of a sub-atomic particle or the body of cultural knowledge contained in a stone-aged tool—may be the most palpable manifestation of reality in the eyes of postmodern scientists. I consider this information-as-the-new-materiality theme throughout this project. Several SFI scientists recommended this book to me, although one wondered aloud what prompted Johnson, a noted *New York Times* science writer, to veer off onto a seemingly mystical tangent.

Several writers have expressed suspicion about the religious and philosophical overtones that seem to emanate from the Santa Fe Institute. For example, philosopher and cognitive studies theorist Daniel Dennett in his book, *Darwin's Dangerous Idea: Evolution and the Meaning of Life*, 1995, takes issue with Stuart Kauffman for the latter's claim that there is more to the origins of life than natural selection. Kauffman's arguments about emerging order may give people the false hope that God lies behind the scenes, Dennett writes (227). Kauffman's highly literary style of writing and his prolific use of metaphor perhaps contribute to the religious overtones that Dennett senses.

Scientific American writer John Horgan in his 1997 popular and provocative book, *The End of Science*, quotes scientists who claim that "complexity" and its cousin term "chaos" have been defined in so many ways that they cease to mean anything. Horgan argues throughout that scientific theories implying an underlying order behind random events are impossible to test because these theories assume a certain level of indeterminacy in the system. Science that cannot be tested or falsified veers perilously close to philosophy, Horgan argues. He suggests further that the SFI type of science is "ironic" in that it blends philosophy and science into a type of speculation about questions that may never be answered (7).

2
A Technical Writer at the Think Tank

I first approached the Santa Fe Institute in the spring of 1997, asking if members had any writing assignments. I was attracted to the science conducted there because it was interdisciplinary and at times seemed almost spiritual in its assumption that underlying patterns govern an apparently random and uncaring world. Like many enthusiasts of popular science writing at the time, I had read several of the books about the Institute that discussed complexity in physical (non-living), natural (living), and human social systems. From reading these books it was obvious that words and their meanings were a source of discussion at the Institute; a writer studying the technical use of language would find much to ponder in terms such as "fitness landscape," "sand pile catastrophe," and "complex adaptive systems." My thesis for a 1995 master's degree in economics looked at chaos theory in production, drawing from some of the SFI research. So I was keen to develop a relationship.

The Institute's then vice president for academic affairs, Erica Jen, and program director, Ginger Richardson, responded, asking for help with specific articles for the Web page about some of the many areas of research at the Institute. They wanted someone to interview scientists and write an overview article that would, in one sitting, provide potential researchers with a general picture of what goes on there. The need for this overview piece became clear when I was talking at lunch one day to a faculty member visiting from Dartmouth College. His board of regents was interested in forming a relationship with the Institute, but he was frustrated because there was no good source of information that tied together everything the Institute does. So he had nothing in layperson's language to show the board.

It was obvious from the Institute's Web pages that each scientist had done a thorough job of writing up his or her specialty, but the writing was targeted very much at an audience of insiders. Throughout the summer of 1997, I spent many hours at the Institute and at home drafting multiple versions of a new overview piece for publication on the Web. In the body of this chapter I summarize the rhetorical challenges I faced in preparing the Web page piece. The goal is to show how the use of metaphor and its harmonics in science are not the idle ruminations of ivory tower language scholars. These language problems are real, lying at the epicenter of any science that is exposing and transforming theory.

The Web page challenges at the SFI mostly involved the choice of appropriate terms and metaphors to describe and explain the illusive complexity science. These were rhetorical challenges. How could a writer choose terms that would categorize the types of research without drawing artificial boundaries around those categories? Which metaphors should the writer employ to accurately describe hidden interactions among agents in a natural system without falsely adorning those descriptions? How could the writer capture the "personality" of the Institute and its scientists (what would be known in rhetorical theory as *ethos*) without overshadowing the facts of their research (what would be known as *logos*)? The task was complicated by my awareness as a scholar of rhetoric of the tension between science in the positivist tradition, which posits the existence of a neutral reality, and postmodern philosophy, which holds that reality is largely constructed subjectively out of words.

How to use language fairly to show reality as it appeared to these complexity scientists became the fundamental question I struggled with as a new technical writer in a science institute. I quickly became convinced that the main challenge in technical writing in science is not found in the nuances of Web design, software documentation, or related topics that occupy so much of the technical writing literature now. The main challenge is an ancient one; it is one of semantics, of choosing the best words to represent and sometimes mold reality.

In the fall of 1997, Richardson also commissioned me to write a piece for the Institute's semi-annual *Bulletin*, which goes out to donors, scientists, and other interested people. That article on the Swarm software program, cited earlier, was the first of several writing projects that fell between the realms of science journalism and technical writing. Richardson offered me another *Bulletin* assignment in

March 1998; I attended a symposium at the Institute with about forty scientists who spent the day discussing the role of scientific models and how well such models can transfer across disciplines. Throughout much of the following month I planned, wrote, and revised the modeling article, while consulting often by telephone and email with Richardson and *Bulletin* editor Lesley King. Later in the spring, however, King notified me that Jen had decided not to publish my piece for reasons not entirely clear. Jen's reasons appeared to have more to do with the rhetorical tone of the piece than with its content. My choice of words and metaphors to represent reality for the first article on the Swarm research was acceptable, but the subsequent effort crossed the line from representation to molding or even adornment of reality. Crossing this line was an unacceptable trespass for a technical writer who would not appear expert enough to produce knowledge.

ESTABLISHING ETHOS WITH AN "INITIAL IMPRESSIONS" REPORT

Before embarking on the Web page overview piece, I spent a few days in the summer of 1997 interviewing several administrators at the Institute, including Jen and Richardson. The purpose was to determine what these insiders thought a general Web page article should accomplish, and what it should contain. I conducted these interviews in a journalistic or ethnographic style, attempting to hold back my own impressions and instead allow ideas to emerge from the people I was interviewing. I then wrote a report titled "Initial Impressions," which summarized and analyzed the comments I had gathered, and suggested a course of action for the Web page lead article. Richardson and Jen praised the initial report; I believe it increased my credibility and ethos as a thoughtful writer who could listen and summarize what he heard.

The Santa Fe Institute at the time of my report was thirteen years old and, as characterized by one member, in the midst of a mid-life crisis. Members told me the program structure was out of date and compartmentalized despite a strong commitment to the interdisciplinary approach. The programs needed some accountability, members said—perhaps more application to the business world—but without compromising the freedom to explore highly abstract and theoretical scientific research. They seemed to believe that written materials about the Institute, such as annual reports, Web page summaries, and the like, were part of the problem. The sense was that if the mission statements, research summaries, and other materials were revised, such revision would assist in providing a new focus for the SFI. Hence, I felt safe in concluding that Institute members placed a high value on the power of writing as a means of clarifying purpose. Rhetoric for SFI members appeared to function not merely as a persuasive or explanatory tool, but as one that actively creates knowledge and understanding. Members exhibited some faith in writing as a way of consensus building and identity formation.

The Institute faced a clear economic challenge, that is, the allocation of scarce resources. Some programs might have to be phased out to allow new ones to grow. One administrator noted that the Institute had approached its maximum size of fifty or so residents. "We have to decide, if we do this, we can't do that," he said. So it became clear that the decision of how to present information in written texts would be tied to decisions about what shape the Institute takes in the future. It followed that rhetoric at the Institute served not just as tool of knowledge production, rhetoric of inquiry, but in the traditional ways as defined by Aristotle.

Rhetoric for Aristotle functioned in one of three branches: *forensic*, where speakers used discourse to evaluate past actions, such as the guilt or innocence of a person in court who had been accused of a crime; *epideictic*, where speakers used words to praise or criticize; and *deliberative*, where speakers used words to exhort people or a governing body into action. Rhetoric at the Santa Fe Institute that helps determine which scientific programs will continue and which will be dropped is functioning deliberatively.

The primary rhetorical challenge I gathered from those initial interviews was that I needed to develop this written material in a way that both accurately presented the Institute to the outside world and provided new motivation internally. But such text must not restrict or dampen the intellectual freedom. The danger of written text is that it can quickly become authoritarian, seeming to carry high import by virtue of its semi-permanence. Such authority would counter the dynamics of interdisciplinary science, where new ideas evolve incessantly to be tested against old ones to form new knowledge. Jen said, "We want to preserve the freedom of the Institute to pursue work that could not be done anywhere else." Any text must avoid a tone of stifling permanence and authority. The notion of evolving thought and of living ideas must be obvious. I suggested in my initial report that maintaining a sense of living ideas would perhaps be easier to accomplish on the Web than in white paper reports; text on the Web can be presented as a fluid "society" of concepts linked by hypertext rather than as a decree cast in stone.

In the "Initial Impressions" report I wrote that the first goal in revising Institute written material would be to prominently display its philosophy of science and the types of research carried out under that vision. The research is most important. Yet, anyone who looked at the Web page in 1997, for example, had to click through several hypertext links before getting to the research. Once there, a reader would find a menu of subtopics that nominally did not appear to have any relationship to each other. Administrators suggested that these reports appeared as compilations of the different branches of research, but not as an integration of that research. Hence, I proposed an early effort to emphasize a common theme underlying the fissiparous research organisms at the Institute.

Implicitly, the search for a unifying theme leads to metaphor because metaphor transports ideas across knowledge fields. The existing SFI term that functioned as a unifying concept appeared to be "complexity." Jen said, "We do have a core set of metaphors that relate to complex adaptive systems." But as I was conducting my research, I picked up some concerns about whether a new unifying term should have emerged by now. Jen and others seemed to view the word "complexity" as a metaphorical term, but they did not clarify whether they saw it as such, or why they would want to have any metaphor at the center of their research. I did note that members were asking whether "complexity" was still the best unifying term for cellular automata, genetic algorithms, a fat-tailed stock market, and similar research. That question continues to nag Institute members.

The immediate rhetorical challenge seemed to be in finding discourse that crosses disciplines, but does not sacrifice too many of the specifics of each discipline so as to render description of these disciplines inexact. Rhetorical theorists might say that a goal of the Institute was to find language that would make various scientific fields "commensurable," although they would acknowledge that such a goal is difficult, if not impossible, to achieve. A danger I was aware of in seeking a unifying theme or meta-discourse was giving the impression the Institute is too reductionist in a search for the "theory of everything." Institute members noted that they have been criticized for suggesting such a sweeping goal in the past and they want to avoid furthering this impression. Perhaps for that reason, I knew I would have to be cautious when writing about metaphors at the Institute or when choosing metaphors to use in my technical writing.

Attempting to find a common language for all branches of science would be as artificial as past efforts at a common language of Esperanto, one administrator said. He suggested that instead of forcing different scientific phenomena to wear the same metaphorical cloak, we should use similes. Similes are less powerful as literary devices, but perhaps more honest. So we might say that a phenomenon exhibited by swarming bees "looks like" the behavior of economic agents—drawing a parallel without applying a metaphor, which implicitly assumes common cause or agency.

But Jen presented a potential benefit from common metaphors. She said that each scientific discipline trains its focus on one level of observed activity, while ignoring others. Such coarse graining is necessary to avoid research paralysis, she said. Chemists, for example, ignore the subatomic level of matter and focus on higher-stage interactions, while physicists focus on the subatomic. A common metaphor, she said, can help researchers from various disciplines remember what they have defined away; it can help them see the big picture. The contrasting opinions that administrators held about metaphor suggested to me then that the power and the dangers of metaphor in SFI science would be a rich topic for further research.

Metaphors are used not only in collating disparate research at the SFI, but also in explaining the amorphous structure of the Institute. Various descriptions I heard or read include "a virtual community" and "a growing extended family of restless minds." It was clear that maintaining this non-hierarchical organization was important to members of the Institute. At the same time, the challenge became that of trying to figure out the relationship between the virtual community and the physical Institute. It appeared then that any organization that downplays physical location becomes more dependent upon other unifying entities, such as a common discourse that is associated with that organization by people in various off-site locations. I sensed that that the metaphors that were known among the scientific and nonscientific outside world as "Santa Fe language" might instantiate a virtual campus that extended well beyond the actual physical site. Figurative language, then, becomes a type of "place," a rhetorical world inhabited by people who speak the same language regardless of where they are physically located.

Yet, the presence of Santa Fe Institute discourse revealed another obvious rhetorical challenge in the management of specialized terms, such as "complexity," "nonlinear systems," and the like. Some of these terms are relatively new to science and their use invites confusion because many have non-scientific meanings. As physicist Murray Gell-Mann pointed out in one lecture, "complexity" to the average person means "something difficult to explain." But at the Institute, the term refers to a self-regulating, self-organizing system whose behavior is emergent. In my initial work at the SFI, I thought that defining terms and distinguishing them from popular usage was necessary to limit confusion. Yet, scientists have no proprietary claim over language, which led me to ask in later interviews with scientists if the Institute would be better off using less harmonic and evocative terms, or even mathematical symbols, to avoid discourse distortion.

Furthermore, as we saw in the first chapter's cursory exploration of some of the popular books about the SFI, the new science of complexity has elicited some skepticism from the larger scientific community outside the Institute; SFI science seems unfalsifiable, and hence, perilously close to philosophy. Yet, scientists and administrators at the Institute seem hostile to the notion that they are practicing a form of transcendental metaphysics. In one email correspondence with Richardson, I stated that work by Stuart Kauffman seemed almost "spiritual" in its assumption that human life on earth was inevitable. "Enough spirituality," Richardson replied tersely.

A TECHNICAL WRITER'S BALANCING ACT: CAPTURE THE EXCITEMENT OF RESEARCH, BUT AVOID HYPE

Administrators were wary of euphoria about the Institute. They seemed united in their feeling that the Institute's reputation has suffered in the past because of excessive hype. Much of it has been generated by the work of perhaps overzealous journalists and book authors who have been giddy with the possibilities of a new science. Popular science writing is replete with terms like "revolutionary new discoveries," "scientific luminaries," "paradigm shifts" and the like, which can cast the scientists and scientific institutions that are subject to this adoration in a grandiose but non-scientific light. "Our main problem has been over-hype," Jen said. "People don't want to be associated with the place—it's too flaky. The criticism is that some researchers are legends in their own minds."

Part of the problem of over-hype is unavoidable. Research in rhetorical theory, particularly the work of Fahnestock in "Accommodating Science," shows that the popular writing of science tends to be epideictic, full of praise and laudatory display. Popular science writers focus on the "wonder" of science and its potential applications. The writing of specialists is more forensic, providing a cautious look at new research that attempts to carefully assess its merits. So where the audience is one of generalists, the epideictic display may be appropriate. An audience of peers, however, requires a flatter style of prose. Defining audience is a significant rhetorical challenge I identified at the Institute, and one that generated some debate.

Jen suggested that all written Web page material be pitched at an audience of potential researchers and post-doctoral students. The writing should be careful and sophisticated and not overstated. Jen did not want to address other audiences, fearing that the more laudatory prose would compromise the Institute and erode its respect. Others, however, suggested that multiple audiences could be considered, not necessarily by adopting different tones, but by presenting different levels of detail. So a Web page report on research in Swarm theory, for example, could summarize the research in lay language in one or two paragraphs. The summary could include a hypertext link to a more detailed, data-filled scientific explanation of the research.

The Institute's home Web page suffered from not having any unifying section that explained in one sweep what is going on at the Institute. Hence, someone arriving at the Web page would have to know in advance what particular branch of research he or she was interested in exploring. Yet, the particular branches of research carried cryptic metaphors such as "ECHO," "cellular automata," "self-induced criticality," and the like, which offered little information to the uninitiated reader about what these branches of Institute science considered. Because the SFI practices non-traditional science, the new visitor would not benefit from the traditional category markers of science—terms such as "biology," "physics," and "economics." It seemed obvious that the Institute needed one summary article built around the idea of complexity; it should describe the various branches of Institute research and how they related to each other. The difficulty would lie in representing this science forensically when its collection of metaphors was so evocative, so sparkling with harmonic associations that clamored for epideictic display.

During the summer of 1997, I interviewed various scientists and read published reports. Then I composed a single piece (about 2,000 words) that tried to organize the Institute's work into a coherent theme. I spent much of my time hewing out the unifying theme of complexity in what journalists often refer to as "a nut graph." Here is what I came up with after several tries and after comments from Richardson and Jen:

> It is difficult to group the research in traditional academic categories or programs when you are in an environment as fluid as the SFI. Because some of the questions being asked here are new and cross many academic disciplines, scientists often find it challenging even to define the concepts they are studying. But it is safe to say that the SFI studies tend to follow living and non-living agents and groups of agents as they emerge, as they organize themselves into complex communities and networks and as they adapt, evolve and learn. The processes of emerging, organizing and evolving often are inseparable; in a way the three are merely different filters through which we view the dynamics of complex systems.

I developed this paragraph during a pencil and paper brainstorming session in which I listed all the research projects extant at the Institute and attempted to discern a relationship or hierarchy among them. I first tried to group the projects according to topics, in which I would include projects related to simple living systems in one group, projects related to more complex living systems in another, and so forth. I quickly realized, though, that such a topical approach would lead back to the traditional categories of science. An epiphany came when I decided not to organize the Web page article by nominative categories based on states of condition, but by predicate categories based on action. Hence, I came up with categories reflecting three stages of development: emerging, organizing, and adapting to changes. Projects dealing with the development of social and economic agents, or of biological cells, for example, could be seen as focusing on the act of *emerging*; projects dealing with the collection of those agents into social groups or groups of cells could be viewed as focusing on the act of *organizing*; and projects dealing with evolution and learning among those groups could be viewed as focusing on the act of *adapting*.

While acknowledging in my Web page article that these classifications were arbitrary, I still believed such classifications were needed in order to develop a coherent overview piece. Without this classification template, the Web overview piece would simply have been a collection of disjointed summary articles about the individual projects. This was the very problem we were hoping to solve. It became clear that the process of classification, while admittedly arbitrary, is essential for human cognition, which requires an individual to recognize patterns and relationships among disparate concepts. Categories are semantic constructs, of course. This insight strongly supports the claim that classifying science is constructing it. Recall the deep philosophical problem that categories raised for Aristotle, Kant, Wittgenstein, and others; those great thinkers questioned whether the categories of knowledge are universal—"transcendental" in Kant's language—or arbitrary. Even new branches of interdisciplinary science cannot escape this need for classification, but at best can hope to forge new categories that more faithfully capture new understandings of the patterns and metaphors of science.

It took six drafts with writing, interviewing, and revising to come up with a Web page report that Richardson and Jen felt offered the right balance of general information and specific scientific precision. My first drafts clearly were too folksy and epideictic ("gee whiz" awe, as Fahnestock might argue). In my conversations with Richardson and Jen, we came to specifically define the audience for the Web page as primarily scientists who might want to visit the Institute to study. So nearly all traces of epideictic popular writing needed to be purged. Here are some examples of what I wrote that did not make the final cut:

> "Some of the scientists who have passed through the Institute have characterized it as fostering a type of 'Renaissance Man' science in the true interdisciplinary tradition of Sir Isaac Newton and Charles Darwin." (*Jen's comment: "You can say that about us, but we can't." She wanted to be sure the Web report did not show signs of hubris.*)

> (*In setting the Institute scene*): "Or you may attend a talk about complexity in information systems by resident physicist

Murray Gell-Mann whose postulation of the subatomic quark particle earned him a Nobel Prize for Physics in 1969." (*Again, Jen wanted to avoid "hype" and the cult of celebrity.*)

Whenever I developed a narrative for the Web page piece suggesting that Institute members were on their way to solving scientific puzzles or producing applicable science, it was cut. Jen and Richardson and the scientists I interviewed did not want to lay claim to any big accomplishments; hence I presented the work as in-progress and not definitive. Compare the initial draft and final Web page sections below about evolutionary biology. A reference to possible applications of the Institute research in "solving mysteries" of Martian rocks was cut. Also, Institute editors softened the dramatic narrative for the final version; references to "this roiling soup" and other poetic metaphors of what happens on the ocean floor were deleted.

Finally, notice that the initial narrative highlights one person, biologist Stuart Kauffman, whom I perceived to be a celebrity because he has several recently published books. In the final draft, I was asked to include the names of several other researchers. The changes again reflected a desire by Institute insiders to avoid a cult of single celebrities and, instead, show a more collaborative approach among various individuals of equal status. Both versions give credit to scientists at other institutions also working on related projects; I had been told to include such acknowledgments to counter the impression of self-absorbed pride among Institute scientists.

One group of researchers at the Institute is asking how the first replicating biological life forms emerged on earth. Drawing on the work from other Institutions, SFI theorists Stuart Kauffman and others are considering how life may have evolved around volcanic vents deep in the ocean. These "smokers" convert the ocean water to a hot organic solvent. But questions remain about how this roiling soup generated molecules that were capable of taking action to maintain themselves. What thermodynamic conditions were necessary for systems to become more ordered and specialized? Answering these questions ultimately could help solve some of the mysteries of the micro-fossils found in Martian rocks, which suggest that primitive life may also have emerged on that planet.

(Author's second draft)

As an example of current work, a group of researchers at the Institute is asking how the first replicating biological life forms emerged on earth. Experimentalists and theorists at SFI and elsewhere—including Harold Morowitz, Stuart Kauffman, Reza Ghaderi, Peter Wills, Philip Anderson—are embarking on a new, joint approach to some of the questions that have plagued the study of the origin of life for decades. What were the first replicating biological molecules on earth? What are the thermodynamic conditions that have to be satisfied for systems to become progressively more ordered and specialized? What sort of chemical reaction network is needed to produce anything as complex as cellular biochemistry?

(Final Web page)

Surprisingly, when the Institute published the Web page piece, it was prefaced with an introductory paragraph describing the author as a journalist who had spent some time at the Institute, and who was now summarizing his insights. A graphic inset on the page included the subheading, "one journalist's impression." I had expected the piece to appear as an "official" Institute summary, since so many people had a hand in the prose it contained. I had expected that my individual ethos, or character, would be subsumed into the larger ethos of the Institute—that I, in effect, would become a member. Perhaps, even after all of the rewrites, Institute decisionmakers still felt the need to distance themselves from anything that carried vestiges of epideictic outsider prose. While I was pleased with the piece overall, and with the public credit I received, I felt a little uneasy with the introduction describing the work as that of a journalist; journalists do not write in conjunction with their sources.

WRITING FOR THE INSTITUTE'S *BULLETIN*: WHAT HAPPENS WHEN TECHNICAL WRITERS CANNOT FIND THEIR BEARINGS TOWARD AN AUDIENCE

I composed the first *Bulletin* article in the fall of 1997 about the Swarm software—a complex systems simulation system—after interviewing the system's founder, Chris Langton, and several users of the system in business and academia. This piece naturally flowed in a magazine journalism style, employing several running anecdotes that appeared throughout the article. One anecdote asked the reader to imagine she was an archaeologist attempting to understand why Native American peoples dispersed from their desert homelands some 1,000 years ago. The Swarm computer simulations allowed researchers to determine that the Native families may have traded corn in times of local drought, but that eventually environmental degradation forced them to move on. In reporting on this research, I seemed to have stumbled upon the right mix of forensic science and epideictic journalism; this article was published virtually unchanged from initial drafts.

The second *Bulletin* article, completed in the spring of 1998, posed many more writing challenges. I had a much more difficult time finding the proper tone for this article and finding my audience. Several explanations for these problems come to mind. First, the article was a report on the Institute's 1998 science symposium on modeling, which drew nearly a dozen speakers from various fields. Their talks did not always seem tightly focused on the subject of modeling. Hence, I was left with the challenge of discerning a common theme, much as I was with the Web page project. But unlike the Web page project, I did not have hundreds of pages of written material to peruse in search of commonality. I had to draw primarily from notes based on short presentations. As a technical writer on the Web page project, I was assembling a superstructure of reality from existing pieces. These included metaphors that had been tacitly admitted into the Institute's lexicon. For this second *Bulletin* article, however, I was being asked to forge and transform reality, which required a technical writer to become a theorist.

Second, I was writing the article on modeling in the midst of my own studies on the rhetoric of science, which offered me new insights, but also diverted my attention from the symposium to wider philosophy of science issues. Because of my studies, I felt much more like an insider writing this new piece than I did writing the Swarm piece. I believed my quasi-insider status would allow me to contribute new knowledge about modeling rather than merely report on existing knowledge. Still, such perceived insider status robbed me of the rhetorical distance (and perspective) from my subject and from my audience. Such perspective had made writing about Swarm much easier.

The first draft titled "Scientific Models at the Santa Fe Institute" in early March, 1998, was almost 7,000 words long and tended toward a blend of academic and journalistic styles. In addition to reporting on the one-day seminar, the article included citations from philosophers of science Max Black and Michel Foucault, and references to Aristotle. Richardson and King sent an email back that the piece was too long and full of digressions. They wanted a piece that was more perky and, to my surprise, more epideictic and journalistic. They did not specify epideictic journalism, but their comments steered me in that direction and reaffirmed my insights that one of the Institute's rhetorical challenges—aside from the semantic challenge of metaphor—was in narrowly defining audience. Readers of the *Bulletin* would want prose that is accessible and that flows, not prose that is turgid and overly affected with a perhaps strained new academic's voice. Yet, writers must not trivialize difficult concepts or "write down" to the audience. Any writing for the *Bulletin* would have to strike a delicate balance between these seemingly contradictory rhetorical demands of audience consideration.

After corresponding with King and Richardson, I determined the audience for the *Bulletin* to be more general than the one for the Web page. This *Bulletin* audience seems to be mainly benefactors rather than potential researchers. Hence, more journalism was necessary to serve the outsiders. Ironically, I seemed to have intuitively known this when writing the Swarm piece, but had to relearn my instincts when writing the modeling piece. Perhaps the process of trying to move from outsider to insider status leads a writer into a state of hesitancy, where he can no longer be sure of his place and of the rhetorical customs appropriate to that place.

The first draft was criticized for not developing a strong argument about what happened at the symposium and for not centering all additional reporting on that argument. Jen and others acknowledged privately that the forum results were unclear and haphazard. Still, I was asked to come up with some common theme. The article's implied thesis was that modeling is difficult across disciplines and that the symposium produced as many answers as questions. I did not mention a common theme emerging, because I did not see one. As an aspiring insider, I was reluctant to make epideictic pronouncements of accomplishment, but preferred now to maintain a forensic tone of hesitancy. But King wanted something more definitive. Perhaps she and others wanted to show results to the *Bulletin* audience, suggesting that the symposium was not a waste of time. Here, then, was the final thesis statement, which King suggested:

Typical for the Santa Fe Institute, as many questions surfaced as did answers; thinkers from various fields pushed their thoughts to the edge of their understanding. Nonetheless, a consensus emerged that models can apply across extremely diverse fields, but not without difficulties.

Whereas I was discouraged from putting exciting narrative into the Web page piece, I was told to add it for this *Bulletin* piece. My first opening sentence, for example, reported matter-of-factly that microbiologist George Oster used engineering terms to describe the action of ATP synthetase in cells. King asked for "a gesture, something so we can see him." Here is my final version:

Some 40 scientists sat in rapt attention as microbiologist George Oster threaded his way to the front of a conference room at the Santa Fe Institute's 1998 Science Board Symposium in early March.

This second *Bulletin* article was never published. After several months of editing and re-writing, at the instruction of Richardson and King, the article went to Jen. She decided not to use it, but to instead work with a graduate resident at the Institute to produce a different account of the annual science board meeting. Jen explained in the summer of 1998 why she chose not to run this piece. "I made the choice," she said. The new piece "was much safer." Jen elaborated: "The danger of antagonizing some of our hard-core scientific readership is greater than boring our non-obtuse scientific readership."

"People were being asked to talk about their work," she continued. "It's risky, very easy to alienate people. What you ended up doing was supplementing and putting into context." Jen said the task that Richardson and King were asking me to accomplish—that of developing a unifying theme for the conference and putting it in understandable language—was not possible. "The raw material didn't have what we were asking you to do," she said.

She suggested that my analysis of the difficulties in developing models that translate across disciplines might have been too candid. None of the speakers addressed the question of interdisciplinary modeling, Jen noted, implying that my effort to do so perhaps would have placed the scientists in the position of appearing more unified then they were. Yet, revealing the failure of this symposium to address the question would imply that for all the talk of interdisciplinarity, these scientists were as narrowly focused as any who are bounded by a single discipline. "For this event, maybe we didn't want to air all of the dirty laundry," she said. "Better for us to err on the side of being stodgy and correct than provocative and qualitative."

In the end, I realized that my initial instinct was probably correct. This meeting did not cohere around an easily summarized theme. Efforts to make it do so required transfusions from philosophical discourse that lay outside the boundaries of the meeting. Jen's instinct was to sacrifice some thematic unity in order to not artificially tidy up a messy picture. It seems clear that efforts to forge thematic unity are driven by rhetorical considerations that are motivated by a literary, or narrative, view of reality—in the sense that human beings like stories in which all the characters relate to one another. Such a quest for thematic unity also is present in science. It can create tensions for scientists who may want to tell a story, but who may never be sure if in doing so, they are manipulating the data to make it cohere. This quest is especially evident when scientists attempt to tell a story of cause and effect. In writing about this symposium perhaps the underlying rhetorical tension was a scientific tension over whether the disparate presentations from various researchers could be linked together to cause a conclusion about the role of models in science.

Of course, it is possible that I began my article with a philosophical approach that lessened my attention to scientific precision. Arguably, the prose appeared to Jen as sloppy science. Jen said that some of my characterizations of scientific concepts such as non-linearity and discrete-versus-continuous time were inaccurate, but that such inaccuracies would have been easy to fix. But then she made an interesting contradiction, suggesting that some of these concepts are nearly impossible to render in lay terminology. "These are complicated enough issues, almost impossible in a sentence to explain without qualifying so much to the point of being meaningless." This comment reveals a profound dilemma that scientists face. How can they use language to convey meaning of what may be an amorphous theory—an impression of reality rather than a representation? The scientist struggling to represent a vague aspect of reality may find some language too precise for evoking impressions. Metaphoric language, with its harmonic overtones, certainly is impressionistic, but perhaps at the cost of not allowing the scientist to control the harmonics. So that scientist might feel that there are two bad options: over specifying a theory with a lot of qualifying language or under specifying it with metaphor. This is the frustration I sensed that Jen was expressing.

Reflecting on these writing experiences at the Santa Fe Institute during 1997 and 1998, I concluded that complexity scientists are highly sensitized to the issues that rhetoricians and philosophers of science discuss. Questions of audience surface often, especially in the pre-writing stages of documents. For the Web page piece, Institute administrators envisioned a scientific audience. Hence, even though I first missed the audience target by writing too epideictically, I quickly recognized my mistake and realized what needed to be done to revise the project to address the specified audience.

The first *Bulletin* article about Swarm was successful because it focused almost exclusively on the work of one scientist, and hence did not require an outsider to synthesize the work of various insiders. Writing as a science journalist, I was able to report on one scientist's knowledge without having to intrude inside his world. In the second *Bulletin* article, however, I found myself having to move beyond what the mere "knowledge telling" role into a "knowledge transforming" role (Bereiter and Scardamalia 3–30); I had to subsume the voices of various scientists into my voice. This required me to function not as a journalist, but as a technical writer whose job was to synthesize theories. Michael Polyani, Thomas Kuhn, and other theorists of language and science have convincingly shown that much scientific knowledge is "tacit," which can be extrapolated to mean that "truth" in an organization like the Santa Fe Institute is found in the unspoken assumptions of insiders. Yet, I lacked fluency with that tacit knowledge and the language—including metaphors—used to create it, and, therefore, did not command the trust afforded to insiders who would judge my writing. The SFI, like many institutions, is reluctant to convey insider status to outsiders, as I also learned when my highly collaborative Web page piece was rhetorically distanced from an insider position.

While the second *Bulletin* article experience was frustrating, it reinforced my insights into the power of metaphor at the SFI. At the heart of all the models discussed were metaphorical statements: cellular metabolism as a mechanical pump; earthquakes as railroad cars that jerk and slip when they begin moving along the track; a business as a human being that can grow old and fat and eventually die of poor health; the economic system in Renaissance Italy as a sand pile that gradually changed one event at a time until one grain of "sand" finally plunged the system into civil disorder—an "avalanche."

The science of complexity is relatively new. Hence, its practitioners are busy trying to define basic concepts such as what is the difference between "emerging" and "adapting" or what is a "rule"? These practitioners almost subconsciously are admitting some metaphors into the lexicon, e.g., "swarm" or "sand pile," while rejecting those they might see as excessively florid or full of poetic harmonics, e.g., "roiling soup." Fine tuning the rhetoric is most essential at this definition stage—perhaps more so than at later stages, when the science matures and moves away from defining and describing and into testing past insights.

Any scientist building expansive theory is forced to make semantic choices; these choices involve metaphoric language replete with harmonics that elaborate the theory and differentiate it from other theories. A technical writer working with scientists at this theory building stage must be highly tuned to a culture that is constantly but tacitly making new metaphoric and rhetorical moves. In this culture of semantic science the purpose of research is to define metaphors and other terms. Realizing this, I was driven to further explore the rhetorical challenges that were unfolding at the Santa Fe Institute. This led to the formal study of metaphor and the other language issues that I had glimpsed while doing technical journalistic writing for the Institute. Before delving into that qualitative study, however, it is important to explore the existing body of research that relates to metaphor and how it functions in science. We turn to that body of metaphor theory in the next chapter.

3
Metaphor: Constituting or Decorating Theory in Science

In some ways it is remarkable that a rhetorician would be welcomed into a science think tank like the Santa Fe Institute, given the often agonistic relationship that has existed on and off for centuries between scientists and scholars of the arts. This antagonism has been especially evident when non-scientist writers have invaded the territory of science in an effort to make its findings accessible—even exciting—to a popular audience. That many SFI scientists talk openly and exhaustively about the role of language and metaphor reveals much about the state of science at the new millennium, a time when science, philosophy, and the language arts perhaps are converging again in the search for useful knowledge. Such openness would have been unusual just a generation ago, and throughout much of the history of post-Enlightenment science.

Scientist C.P. Snow bore witness to this conflict in his now-famous 1959 lecture in which he bemoaned a world of "two cultures." One, a world of literary intellectuals, and the other, of scientists, were separated by smoldering suspicion and misunderstanding, Snow said. "There seems to be no place where the cultures meet" (17). When modern scientists accepted a role for rhetoric, they seemed to do so only after the fact, as a way of presenting knowledge already gathered by a non-rhetorical scientific method. Dorothy Winsor notes in her research ("Invention and Writing" 227) that engineers often talk about "writing up" their work as if science preceded writing, but did not accompany it. For such modern scientists, the role of language was to describe nature transparently. This type of writing is known as "instrumental" writing, a tool for displaying a predetermined reality. As geologist and science writer Scott Montgomery notes in *The*

Scientific Voice, the modern scientist assumed "that language can be made a form of technology, a device able to contain and transfer knowledge *without touching it* (2, italics original).

Science in Snow's polarized world is a precise culture of facts, observation, and data, while literature would seem to be a fuzzy culture of lore, interpretation, and words. This modern science emerged during the ages of the Renaissance and Enlightenment, at a new dawn of human understanding, when reason and observation were joined together to sweep aside old mythological accounts of reality and replace them with careful hypotheses. It is undeniable that a great awakening began in the sixteenth century that liberated humanity from a dark intellectual period of history. Understandably, scientists mistrusted any impulse that appeared to arise from those darker currents. Language—the essence of what it means to be human—when used hastily, unaccompanied by reason, could be seen as such an impulse. Yet, it is important to realize that science and human narratives were partners for thousands of years, and it is just in the past 400 or so that they have been separated. To some extent, the current movement toward rhetorical studies of science and the rapprochement between science and the language arts is recognition that science and the human experience cannot be detached.

The degree to which science and language are intertwined is by no means resolved, however, and it is the subject of much scholarly research in a branch of rhetoric studies known as "the rhetoric of science." Some twentieth century rhetoricians have argued that science is nothing more than human storytelling and rhetoric, or, as rhetorician Alan Gross claims, that science is "rhetoric without remainder" (quoted in Gross and Keith, "Introduction" 6). Others, such as Dilip Gaonkar, counter that science is different from rhetoric, and that attempts to blur the distinction serve only to weaken the power of rhetoric by, in effect, forcing rhetoric to overstep its bounds. Gaonkar's argument suggests that rhetoricians may sacrifice rhetoric's tradition as a practical art necessary for public decisionmaking when they try to make it a tool of scientific analysis (76).

Although claims that science is a subset of rhetoric seem exaggerated and even specious, all technical writers schooled in rhetorical theory owe a debt to Gross and other such scholars for launching the rhetoric of science program. This program gives technical writers a purchase in science; it allows second-generation scholars to make claims—as I am doing here—about the power of metaphor and other tropes in constituting scientific knowledge. The rhetoric of science program, in turn, is indebted to sociologists and philosophers of science who have helped to lift the veil of objective infallibility from science. Of course, it is no surprise that scholars from these disciplines have disagreed about the ways in which science is subjective. Sociologists of science have argued that the act of accruing and interpreting data is largely a social process by which knowledge is determined to be legitimate among a community of scientists. Those sociologists committed to a "Strong Programme" (associated with the Edinburgh school) argue that scientific truth is meaningful only locally among a society of scientists (Longino 18). Hence, for these sociologists developing scientific knowledge is a challenge of gaining acceptance for ideas within a larger group. For many philosophers of science, however, knowledge making is a cognitive problem of developing theories and representations of reality that are rational, logical, and compatible with observation. Social forces certainly influence this cognitive process, but ultimately the nature of reality will drive theorists to accurate representations that transcend social idiosyncrasies.

Rhetoricians also fall into one of several camps when considering the role of discourse and rhetoric in science, as stated by rhetorician David Fleming (personal communication, 1998). At one extreme are those scientists who see language as irrelevant to true science. In the middle are theorists of the "weak program" of the rhetoric of science, who argue that science and rhetoric converge when scientific data is written up or otherwise presented to the lay public. A scientist who believes that language is important in helping to secure grant money would ascribe to the weak program that Fleming outlines. Proponents of the other extreme, the "strong program," like Gross, argue that rhetoric produces science; discourse is the sole epistemic of scientific knowledge. Hence, the only way we know anything is through language. Thus, the rhetoric of science becomes a rhetoric of inquiry. For proponents of the strong program, "science is just another political debate" (Fleming, personal communication). They would argue that science is as much about the hegemony of power as it is about the production of knowledge.

Science as a source of knowledge dates to the ancient Sumerians and Babylonians, who attempted to discern order in the night sky. For those ancient people and for the Greeks and Romans who followed, science was never a system separated from the myths and legends of human experience—or from words. Science for the ancient Greeks and Romans was deductive, based on a logical thought process that led from premise to conclusion. It was based on reflections on the nature of the world, but without the intervention of formal study. Hence, the poet-philosopher Lucretius (99–55 B.C.E.) could deduce, for example, that because the world consists of mortal embodied creatures, the world's physical components—earth, air, water, fire—must also be subject to the laws that govern a living body. Logically, for Lucretius, if creatures populating the world will die, so too will the physical world of heaven and earth come to an end, "as certainly as ever they once began" (166). Words were the tools of this ancient deductive logic and, therefore, of all ancient knowledge production.

Yet, rhetoricians of science must be careful to recognize that for the ancients, words operated in different and incommensurable domains. Scientists, philosophers, and the lawmakers of the Greek *polis* were ruled not by one discourse, but by discourses assumed to be functioning under different systems of thought. Aristotle defined science as an "analytic" system that adduces demonstrable facts. Science was a deductive process that began with premises and then followed an impersonal system of logic to draw irrefutable conclusions. While Aristotle did make observations of marine life in the Aegean Sea, his science was largely based on mental exercises. Aristotle's science contrasted with "dialectic," a type of philosophical argumentation aimed a universal truths, which Plato called "Forms." The domain of the legislature was left to rhetoric, a type of popular argumentation that attempts to prescribe a prudent course of action (Greek; *phronesis*) as agreed among concerned parties who have assessed the probable state of reality. In Aristotle's worldview, rhetoric dealt with the concerns of people, while science dealt with the facts of nature.

Most scholars of rhetoric today would reject Aristotle's distinction between the discourses of science, philosophy, and rhetoric and argue that once words are compiled into purposeful sentences for the benefit of an audience, they become rhetorical. Aristotle's distinction between demonstrable facts and rhetoric seems especially arbitrary when we examine ancient literary science writing, such as that exemplified by Lucretius. Although science in the Middle Ages became more applied and perhaps less beholden to deductive philosophy, it was no more scientific in the modern sense than was the science of the ancients. Alchemists attempted to make gold and perform similar supernatural acts of pseudo-science; these acts seem to have been rooted in a Platonic assumption that each element of matter had intrinsic attributes, which distinguished it from other elements.

But later Enlightenment scientists, such as Robert Boyle and Francis Bacon, moved away from such mysticism into a more exacting world, where a mechanistic nature was studied. Sociologist Steven Shapin writes that this exacting science was in keeping with the religious belief system of the day, which saw God as a rational clock maker rather than as an occult alchemist (*The Scientific Revolution* 44). As science moved away from the world of the occult and into the rational, it also shifted from a deductive process to one based on inductive observation. Bacon led the call for experiment and observation, while Boyle began conducting experiments with air pumps and the like that could be repeated by other scientists. Scientists came to value research that could be verified and challenged.

Enlightenment philosophers lent credence to the new scientific method of observation that was unfolding. John Locke rejected the notion of innate truths and Platonic forms and "put experience, or ideas of sensation and reflection, firmly at the basis of human understanding" (Blackburn 220). Rene Descartes argued that true knowledge must be tested by doubt. He wrote: I will "include nothing in my conclusions unless it presented itself so clearly and distinctly to my mind that there was no reason or occasion to doubt it" ("Discourse on Method" 15). In the early nineteenth century, Auguste Comte distinguished three phases of human understanding and argued for a "positive philosophy" that would reveal the workings of nature. Comte proclaimed that human understanding of nature moves through various stages—a process in which reality is first seen as the result of life-like transcendental *forces* and finally, as the product of almost mechanical *laws*. For Comte, human understanding first resided in a theological state, where natural phenomena were seen to be caused by supernatural beings, or deities. In the next state—the metaphysical—supernatural forces were replaced by abstract forces of nature, where nature almost assumed the power of a deity. In the final positive stage, the search for life-like forces gives way to a search for the laws that connect observed phenomena to statements of facts.

This doctrine was refined in the early twentiethth century by Austrian philosophers of the Vienna circle, who argued from a position of "logical positivism" that scientific statements must be verifiable. Karl Popper laid challenge to the verification principle, however, revealing that some statements that would seem to be true could never be verified. For example, one could make the claim that all ravens are black, yet not be able to verify it without having seen every raven in the world—an impossible feat (Edmonds and Eidinow 170). Popper agreed that that scientists should posit laws that govern observed phenomena and then test those laws, but the test would come in offering up the statements for falsification rather than verification. A statement that all ravens are black would hold so long as no one saw a raven of a different color. Popper wrote: "Every genuine *test* of a theory is an attempt to falsify it, or refute it" ("Conjectures" 36). Only those theories that could be proven incorrect, then, would meet Popper's falsification requirement. Scientists espousing today what we might call a post-positivist methodology generally are offering a modified version of positivism. Although born out of the positivist empirical tradition, it takes into account Popper's critique. Philosophers of science make the careful distinction between the pre- and post-Popperian methods. (The latter are known formally as "hypothetico-deductive methods" because they deduce logical consequences from a hypothesis and then empirically test to see if those consequences occur.) For scientists, the main working criterion of the post-positivist method is that hypotheses be testable in some way, preferably by observable and quantifiable techniques.

Scientists came to distrust that which they could not see or test empirically, and at the same time, to distrust the mere products of human contemplation. For example, some scientists distrusted Sir Isaac Newton's conjectures about gravity because his gravity was said to act over bodies across a distance, which implied some kind of hidden or occult force (Shapin 64). Yet, Newton's understanding of gravity, while not verifiable empirically, was based upon his own reason. Immanuel Kant helped to bridge the gap between knowledge gained by reason and knowledge gained by observation in his *Critique of Pure Reason*. It argued that empirical knowledge is organized by categories in the mind, such as space and time, which precede human experience. Such categories, however, are not intrinsic to reality; they are frameworks imposed by humans upon their perceptions.

It would be oversimplifying the study of language and science to assume that once experimentation took hold, the power of philosophical and figurative discourse disappeared. Psychologists Dedre Gentner and Michael Jeziorksi show that the alchemists combined a passion for experimenting with various metals with a passion for describing those metals metaphorically, alternatively, for example, as "fiery, active, and male" or "watery, passive, and female" (463). Montgomery, citing Shapin, notes that Boyle, Bacon, and Galileo were expressive writers. Boyle's works on chemistry employed analogy and other figures of speech to highlight his experiments, in an effort to transport the readers as "virtual witnesses" to the experiment site (93–94). Bacon called for a scientific discourse as "a probing and recording instrument." It should not be overly elaborate, but at the same time, must not "be dull or devoid of pleasure" (84). Galileo reported on his observations with the telescope in a language of personal narrative and unabashed wonderment (89).

Still, a change was afoot in science. No longer dependent solely upon imagination and deductive thought, scientists now had the results of observation and experimentation to temper their discourse. Empiricism acted as a kind of check on rhetorical excesses. Bacon warned in his treatise on knowledge, *Novum Organum*, against the idols of the human mind, which would result from dependence upon rhetoric: "It cannot be that axioms established by argumentation should avail for the discovery of new works; since the subtlety of nature is greater many times over than the subtlety of argument" (Bacon 84). For Bacon, language possessed two faults: "These were the ability of a word to possess more than one meaning and the use of—or rather the inability to avoid—metaphor and other figures of speech" (Montgomery 85).

Indeed, it was the inevitable presence of harmonics in figurative language that helped move Enlightenment scientists toward a more mathematical representation of cause and effect. Newton's theory of gravity relied on expressive terms like "attraction." Such terms troubled his contemporaries, who resisted the harmonic implication that animistic desires superceded rote mechanics to influence the behavior of matter. The risk for the early eighteenth century mathematician Joseph Saurin was that theories resonating with such harmonics would return science to an Aristotelian philosophy that saw matter as acting out the hidden impulses of essential qualities. Saurin and many contemporaries preferred to see forces as mathematical abstractions rather than as the result of innate qualities of matter. He feared a return to Aristotle's "old peripatetic darkness, from which Heaven preserve us" (quoted in Heilbron 50).

In the old science of Plato and Aristotle's world, the existence of ideas was necessary before evidence of those ideas could be observed in the world (Reichenbach 22). The new Enlightenment science, however, would suggest that without prior observation, deductive reasoning was merely analyzing reality, but without adding any additional observed knowledge about that reality. Bacon extolled the new empiricists as being "like the bees that gather material and digest it, adding to it from their own substance and thus creating a product of higher quality" (Reichenbach 80). He was speaking metaphorically about a new system of thought that constructed explanations in the context of observation (100). While Hume and other philosophers would mount challenges to the power of empiricism by noting that all observation is theory laden, the practice of science had changed forever from pure contemplation to empirically tempered contemplation.

The very notion of what constituted a scientific experience changed during the period between 1665 and 1800. Rhetorician Charles Bazerman studied papers published by the Royal Society of London and found that in the early volumes, only a few articles were devoted to reporting on experiments. "The most articles and pages were devoted to observations and reports of natural events, ranging from remarkable fetuses and earthquakes, through astronomical sightings, anatomical dissections, and microscopic observations" ("Reporting" 174). Later publications showed an emphasis not on random observation, but upon purposeful experimentation designed to test hypotheses about reality. Formal research became more methodological, based less on qualitative observation, and based more on quantitative proofs. The new underlying assumption was that empirical studies involving observation, description, and analysis of correlation between phenomena have the power to support, expand, or refute theory.

Science was no longer the philosophical contemplation of human beings in the world, but a methodological testing of hypotheses generated by detached observers standing apart from that world. These observers attempted to isolate individual components of that world. Twentieth century philosopher-historian Michel Foucault argues that science placed an arbitrary order upon the world by classifying and representing reality through such new sciences as biology, philology, and political economy (*Order of Things*). This process for Foucault meant, "life appeared as the effect of a patterning process—a mere classifying boundary" (268). Clearly, the world of scientific method, classification, and experiment left a reduced role for rhetoric. Rhetoric could not classify and represent the way science did, but it looked at the world as a series of interrelated problems that mattered to an audience.

The rhetoric-science split remains today, as Snow revealed, although some changes in the mid-twentieth century in both science and rhetoric are bringing the two closer together. Albert Einstein's relativity theory and theories of quantum physics both evoked language that showed the world to be far less certain and logically mechanistic than it had seemed to a previous generation of scientists. Einstein, in addition to being the twentieth century's foremost scientist, also wrote philosophy, and he clearly used writing to help formulate his thought experiments (Minor). Rhetoricians seized upon concepts such as time-space relativity and the wave-particle duality of light to argue that appearances could be deceptive and, therefore, dependent upon the role of the observer. Any rhetorician today who makes a claim on science invariably finds himself reciting the relativist's mantra, borne out of Heisenberg's uncertainty principle, which states that the very process of observing an electron alters the position of that electron. Problems of science became problems of understanding given the limitations of language. For example, as science journalist George Johnson reveals, if we require two seemingly contradictory terms— particle and wave—to describe light, then perhaps the best we can do in describing quantum reality is to "triangulate" that reality using "complementary pairs of imperfect concepts . . . " (*Fire* 147). Far from being certain, science, including that practiced at the Santa Fe Institute, now seems to be about positing probabilities, which are a mathematical depiction of reality, but which seem similar to the world of probable courses of action that rhetoricians inhabit.

The new scholars of the rhetoric of science first looked to Charles Darwin's *The Origin of Species* for evidence that science is shot through with rhetoric. Darwin's nineteenth century masterpiece arguably is the most significant work in science; it also is one of the most rhetorical. In his concluding chapter, Darwin acknowledges that the work is "one long argument" (435). In fact it is a forensic argument designed to build a case for natural selection by arguing that past events could only have unfolded in a certain way. Robert Young argues that natural selection is a metaphor that implies some kind of anthropomorphic selection process ("Darwin's Metaphor" 92). Darwin had to be careful not to alienate religious leaders of his day; he exercised "massive restraint," according to John Campbell ("Charles Darwin" 10). Through much of the work Darwin is forced to consider a priori all sorts of challenges to his admittedly sketchy reconstruction of the past.

Similar analysis revealed the presence of rhetoric throughout science and, by association, elevated the status of rhetoric throughout the second half of the twentieth century. During the mid-1950s, rhetorical theory enjoyed a rebirth that rescued it from the dormancy in which it had been cast in the nineteenth century, when rhetoric had become a mere synonym for the art of oratory and elocution. The key year was 1958, when Chaim Perelman and Lucie Olbrechts-Tyteca published *The New Rhetoric* and Stephen Toulmin published *The Uses of Argument*. These books were enormously influential in recovering rhetoric as a serious knowledge-producing endeavor. Perelman and Olbrechts-Tyteca attacked the elevated status of facts, suggesting that facts are not the goals of intellectual pursuits, but merely the foundations for arguments that concern the universal human audience. They wrote, "A fact loses its status as soon as it is no longer used as a possible starting point, but as the conclusion of an argumentation" (68). The modernist assumption that scientific facts stand above human intervention was challenged in the groundbreaking sociological/anthropological work of Bruno Latour and Steve Woolgar. Their research into a medical biology laboratory in the 1970s found that "facts" are constructed in various stages of certainty, from highly tentative and qualified claims to those statements accepted as common knowledge among scientists. Toulmin's work restored deductive reasoning to the purview of rhetoric by carefully explaining the workings of syllogistic arguments. Toulmin lent a degree of scientific formality to rhetoric, but he also showed that all arguments—even scientific ones—are built upon hidden warrants shared by members of a specific discourse field.

Philosophers of science also have chipped away at the idea of science as an objective and immutable reality. Even basic concepts like "time" proved to be intractable by the scientific method, as eloquently argued by science philosopher Hans Reichenbach. He showed that we can only apprehend time locally, when we can see one event precede another (144–156). Time cannot be known, but only defined by a series of discrete cause-and-effect events. If a concept as fundamental as time is dependent upon definition, then perhaps all the planks of reality are contingent upon defining language.

Such a conclusion lies at the center of Thomas Kuhn's often-cited study of the way science is conducted. *The Structure of Scientific*

Revolutions, published in 1962, argues that real breakthroughs in scientific knowledge (commonly known now as "paradigm shifts") occur in moments of crisis, when existing theories cannot explain observed scientific anomalies. The example he develops occurred in the 1700s, when oxygen was discovered as a combustible fuel, replacing earlier notions that fire was an actual entity consisting of a substance known as "phlogiston" (53–54). A more modern example of a moment of crisis in the social sciences, not offered by Kuhn, could be found in the early part of the twentieth century, when global depression ushered in the revolutionary new economic theories of John Maynard Keynes. Kuhn became the hero of rhetoricians with statements suggesting that language and philosophical theory preceded fact instead of the other way around: "It is, I think, particularly in periods of acknowledged crisis that scientists have turned to philosophical analysis as a device for unlocking the riddles of their field" (88).

Such claims directly challenged the methodology that grew out of the positivist tradition by suggesting that scientific discovery relies on chance, serendipity, and what Polanyi referred to as "tacit" knowledge. Alternatively, breakthroughs in science may happen because of a need for them to happen; in other words, the confluence of world events and exigencies creates some kind of crisis that occasions revision. Certainly, for example, revolutionary theories of twentieth century particle physics emerged primarily because the United States needed a new type of bomb to end its war with Japan. Science philosopher Paul Feyerabend led an all-out assault on rational method in his book title *Against Method*, in which he makes the radical assertion that, "The only principle that does not inhibit progress is: anything goes" (10, italics original). At the heart of this "strong program" challenge to the positivist tradition is the notion that truth is a relative concept that is bound up in what Richard Rorty, the twentieth century pragmatist philosopher, refers to as "individual and cultural webs of belief" ("Science as Solidarity" 44). Rorty challenges scientists to trade in their obsession with rigorous method in exchange for original thinking (51).

RHETORIC AT THE SFI: A QUALIFIED ACCEPTANCE

It seems clear even without extensive research that scientists at the Santa Fe Institute and at similar postmodern research sites are growing more receptive to a renewed role for rhetoric than were their modernist predecessors. Harold Morowitz, a biologist affiliated with the Institute and publisher of its journal, *Complexity*, notes in a 1998 issue that the second half of the twentieth century has seen an explosion of informational sciences. "Of the many papers that cross the desks of the editors of *Complexity*, it has been possible roughly to classify the uses of the vocabulary of the information disciplines into the metaphysical, the metaphorical, the meta-metaphorical, and the magical," he writes (19). Metaphysical science, Morowitz adds, is an Aristotelian concept that relates the dynamics of physical laws to "our understanding of physics, which reaches into the cognitive" (19). Morowitz writes that the use of metaphors is new to science, although he probably is referring specifically to acceptance of metaphor within science in the modern period. This new foray into metaphorical theorizing represents a confidence among scientists to accept theories that arrive from outside the traditional disciplines of classical physics, Morowitz continues. "Thirty to fifty years ago, it would not have been respectable to use the word 'metaphor' in a physics department seminar. Metaphors were for the English department" (20).

Although Morowitz's essay shows just how far SFI scientists might be willing to go in accepting rhetoric, his ruminations suggest that most would be dismissive of the "strong program" claims described above. Morowitz argues that metaphor studies in science have brought on "meta-metaphors," which he holds to be more grandiose metaphors that attempt to link different domains of science, but which are not "used with the same precision" as metaphors, and which are not ground in the rigor of Aristotelian metaphysics. He then implies without stating that the term "chaos" might be just such a meta-metaphor. "As metaphor moves into meta-metaphor, I begin to get nervous about including it, because verification and falsification become much more difficult, and the evaluation of theories tends to become hazy" (20). Morowitz is more dismissive of what he refers to as "word magic" in science, where words are used "somewhat willy nilly" for purely literary reasons (20).

Clearly, empiricism and method still dominate postmodern science; any theory that would attempt to return science to a world ruled by philosophical digressions is unlikely to receive serious attention. Language is a tool in the search for knowledge, but not the only tool. Bacon's exhortations against the idols of excessive philosophizing remain a potent reminder that even in postmodern thought-experi-

ment science, a researcher still has to observe some world phenomena—even if those phenomena are generated on a computer. Still,
these new scientists seem to recognize that despite the warnings of
Bacon and other Enlightenment pre-modernists, science cannot
escape discourse. The challenge, then, lies in using language with care
in ways to create knowledge that is not distorted. This is a challenge
that demands careful analysis of the way words both represent and
shape reality. It is a challenge of rhetoric, particularly metaphor, a subject to which we turn our attention now.

THE MEANING OF METAPHOR

Although poets, philosophers, and scientists have always used
metaphor, scholars have also debated what it is that they are using,
and why. The Roman handbook on discourse, *Rhetorica ad
Herennium*, states: "Metaphor occurs when a word applying to one
thing is transferred to another, because the similarity seems to justify
this transference" (278). Other sources define metaphor as a kind of
simile, where dissimilar things are made to appear alike, but where the
words "as" or "like" are omitted. These succinct definitions, however,
mask the philosophical conundrums that metaphor entails. Words
like "figure of speech," "likeness," "idea," "dissimilar," and "analogy"
are more complicated than they appear. In saying that the "death is
darkness," for instance, are we drawing attention to a true likeness
(i.e., that when a person is dead, visual sensation ceases), or are we
creating an apparent likeness out of dissimilar things (i.e., the absence
of life and the absence of light)? The usual definitions of metaphor
work, but they do not resolve questions over what a metaphor is or
does. Perhaps Kenneth Burke's sparse description of metaphor as one
of the "master tropes" of language—"a device for seeing something in
terms of something else," (quoted in Johnson-Sheehan 48) is the most
practical definition; it does not introduce a lot of problematic terms
in an effort to define a problematic term.

The debate over the definition of a metaphor and its function
lies at the heart of metaphor studies. In asking what a metaphor is or
what it is supposed to do, inquisitive scholars of language are asking
how knowledge is shaped when images from one domain of observed
reality are used to give form and substance to images in another
domain. The word "shaped" in the last sentence is important; by

asking how knowledge is shaped by metaphor (an expression that itself is metaphoric) we are asking if shaping knowledge means creating it or simply altering its form. Using Plato's image, one could envision reality as a piece of clay and metaphor as the hands that mold it. Once molded, it looks different, but it is still clay. So is it new, or does it just appear to be new?

In other words, does metaphor produce knowledge or simply decorate and deliver it? It should unfold throughout this discussion that I have come to believe whenever metaphor is used thoughtfully, it has some role in producing knowledge. Here I am laying aside the age-old debate about whether production of knowledge is independent of its discovery. This debate asks whether science *discovers* intrinsic facts about the world or whether, through the process of scientific discourse, it *creates* those facts. Production in the sense I am using it is a process whereby discovery (the act of observation) occurs concurrently with creation (the act of reflection upon what is observed) to make knowledge. I am arguing that observation and reflection are inseparable; the fusion of the two processes is what produces knowledge, and metaphor assists in that fusion.

Metaphor is one of several tools for producing theory, along with mathematics, empirical research, and intuition. Some metaphors, of course, are more powerful than others. A metaphor that portrays atoms linked together like Tinkertoys to form molecules is naïve, perhaps more appropriate as a means of delivering knowledge to children than as an aid to adult scientific cognition. But almost any metaphor that a scientist reaches for in an attempt to understand reality will have some value in bringing that reality to light, even if the metaphor is later discarded. (Even the Tinkertoy model would have been useful when molecular theory was first developing.) Coming to peace with metaphor's role in producing knowledge is something of importance to members of the Santa Fe Institute. If metaphor produces knowledge, then it is a semantic phenomenon that demands much more attention among scientists and technical writers of science than if it merely describes, delivers, or decorates knowledge.

Debate over where metaphor does its work takes on various guises in the literature of metaphor studies, but much of the debate can be seen in context of this knowledge production-knowledge delivery dichotomy. Metaphor scholar Andrew Ortony refers to the two positions on metaphor as "constructivism" and "non-constructivism," where the former holds metaphor "as an essential characteristic of the creativity of language" and the latter as "deviant and parasitic on normal usage" ("Metaphor" 2). For example, if a nomad says, while hiking through the desert, "The sun today is a raging furnace," he would be making a claim as a constructivist because that he knows the sun's heat on some occasions can transform the nature of the world in the same way a furnace transforms metals. As a non-constructivist, he simply would be saying that the sun today makes him think of a furnace, but that he knows the sun intrinsically to be no different today than he knew it to be yesterday. Metaphor for non-constructivists becomes a means of interpreting reality in a poetic way (where interpretation is non-transformative) rather than as a means of rediscovering it.

Of course, one could argue that a metaphor of the sun as a furnace today hardly advances the science of solar energy, even for nomads. But it is not far-fetched to think that thousands of years ago in the Iron Age, this metaphor—arrived at by contemplation, association with a known technology, and serendipity—may have indeed led some ancient wandering philosopher-scientist to realize that the sun was not a winged chariot of fire, but something intensely transforming, seemingly industrial. The same metaphor that now would seem to be figurative might at one time have been a useful bridge to new theoretical understanding of reality, whereby anthropomorphic images of a flaming God-driven chariot give way to more accurate images of the sun as an inanimate, searing celestial body.

Related to the question of constructivism versus non-constructivism is that which Ortony asks: Is metaphor purely a linguistic phenomenon or more of a communication phenomenon ("Metaphor" 9). As a linguistic phenomenon, metaphor would be integral to the cognitive function of individual words, an assumption that must denote metaphor as inseparable from the bond of signifier to signified. The signifier "sun" points to our signified concept of an object in the sky that, among other things, is a furnace. If we counter that the sun could not be a furnace because it is not used by humans to smelt metal, the metaphor linguist (a type of constructivist) would retort that we must then expand our definition of "furnace" to allow for non-human heating events. The metaphor—more aptly called a "metaphoric statement"—would be transforming the meaning of both terms. As a communicative (non-constructivist) phenomenon, the furnace metaphor would influence the meaning of the word "sun" in a limited poetic sentence-level sense. But presumably after the poetry is finished, we could return to some literal and unrelated concept of the two terms.

A key but recondite concept here is the word "literal." Those non-constructivist theorists who see a metaphoric statement as a relationship between the literal and the figurative assume that literalness exists and that words in normal usage convey that literalness. So an act of metaphor is the use of words abnormally, for effect. Aristotle draws this distinction between the literal and metaphorical in the *Poetics*, where he writes that every word "is either authoritative, foreign, metaphorical, ornamental, made up, extended, contracted, or altered" (39). Language philosopher Paul Ricoeur notes that in this passage, Aristotle links metaphor theory to the larger theory of discourse by means of the noun. An "authoritative" noun is what Ricoeur gives to be "the ordinary word for the thing" (16). The assumption, which shows up throughout analytic philosophy and structural linguistics, is that words represent or stand for components of reality. This is not a trivial assumption; as philosopher Arthur Danto writes, the triangular relationship between the subject and the world and representations of that world forms the foundation of all epistemologically based philosophical inquiry. The early writings of Ludwig Wittgenstein capture an uncomplicated view of the representational component of this triangle, in which "a name refers to an object; it is a proxy in the proposition for the object" (Phillips 22).

In a world of literal representation, metaphor is a literary affectation—an aberrant, ornamental word use. I.A. Richards begins his famous 1936 lecture on metaphor (published 1965) by quoting Aristotle, who in the *Poetics* argued that metaphor requires an eye for resemblances, which "is the mark of genius" (quoted in Richards 89). Metaphor theorists following this tradition have adopted the views that metaphor functions either as a "substitution" of the figurative for the literal, or as an "elliptic simile"—a "comparison" of the figurative and the literal (Black "More" 28). Cicero's definition of metaphor captures the idea that metaphor is a literary ornament employed mainly by skilled wordsmiths to replace literal meaning with figurative. Metaphor for Cicero in the *Orator* is "used for the sake of charm (*suavitas*) or because of the lack (*inopia*) of a proper word" (quoted in *Rhetorica ad Herennium* footnote 278). Metaphor theorist Max Black, however, shows these substitution and comparative views to be dismissive of metaphor's cognitive powers: "The reader will notice that both of these views treat metaphors as *unemphatic*, in my terminology—in principle, expendable, if one disregards the incidental pleasures of stating figuratively what might just as well have been said literally" ("More" 28, italics original). Ortony summarizes Aristotle's comparison view of metaphor as one that assumes metaphors "are not necessary, they are just nice" ("Metaphor" 3).

Yet, literalness unravels before our probing minds when we are forced to ask if a substance is not merely a collection of qualities or attributes. In other words, we can ask as Kant did, if there is a "noumenon," a thing in itself—a literal object or phenomenon that could be called "a fact," or if facts are just the subjective products of human observation? Danto notes that observable attributes can change over time, as Descartes revealed in an example of a piece of wax that melts (Danto 210). In a Platonic sense, then, we could never truly represent an object, but only its transitory qualities. Its true form would remain occluded. Aristotle adheres to the Platonic tradition in accepting that the naming of a thing, or "subject," is different from its intrinsic nature, or "substance." He writes in *Categories*: "Substance, in the truest and primary and most definite sense of the word, is that which is neither predictable of a subject nor present in a subject; for instance, the individual man or horse" (6). From this we ascertain that substances come before subjects, that subjects give crude and inexact form to substance.

Words as subjects in this sense are "ostensive," a term that metaphor theorist Richard Boyd uses to suggest that all representation is a process of naming things according to the way they strike our senses (368). For Enlightenment philosopher John Locke, a concept like ostensive representation would have meant, for example, that in naming a metal "gold," the namer would not be capturing the literal essence of the substance, but only a "nominal essence." These would include the attributes of color, hardness, and so on that relate to gold (quoted in Boyd 364–365). Attempts to reach the literal essence of a term are fruitless, Locke held, because they entail circular arguments. "What is motion?" Locke asked (quoted in de Man 15). It can be defined as a "passage" from one point to another, but then is not the idea of "passage" the same as the idea of "motion"?

It is easy to see where the conundrum about literalness leads in our analysis of metaphor. If we ask what a metaphor is, implicitly we also are asking what a literal statement might be in order to define figurative speech by comparison to literal speech. If the word "complexity" as used at the Santa Fe Institute is a metaphor, what is the literal, substantive concept that it is substituting for, or being compared with? We can create yet another dichotomy and view language from one of two extreme positions—that all language is literal, or all language is metaphoric—or from some point in between those extremes. Analytical philosopher Donald Davidson makes a powerful argument at the literal extreme, suggesting that language only appears figurative. He argues that this figurative appearance is an illusion caused by the ability of humans using language to constantly accept new meanings for old words as necessary. He writes that, "metaphors mean what the words, in their most literal interpretation, mean, and nothing more" (30). So if we read in Genesis " 'the spirit of God moved upon the face of the waters,' " Davidson writes, then we might be tempted to assume that the word "face" has a figurative meaning. But such a figurative interpretation is not necessary, he argues, because the example simply shows a case where the concept of face has been extended to that of water (32). In other words, Davidson might ask, "Who says that water cannot have a face?" For Davidson, metaphor contains no magical power; it is simply a type of simile that "nudges us into noting" similarities between phenomena (36).

The opposite position follows from Locke's claim that words lead to circular arguments, a position taken up by Poststructuralist philosophers who argue that any attempt to attach a signifier to a mental image of an object or phenomena is arbitrary. So, as a Poststructuralist, if I hear the passage about a face on the water, it might conjure up images of one kind of human face—perhaps that of bearded mariner. These images then conjure up all kinds of associations I have about that mariner whose face I envision, yet these associations are so idiosyncratic (and specific to me) that they lead me to a different interpretation of the original passage than you might have been led to, assuming you had envisioned a different face. If words cannot stand for objects or actions literally, but they merely conjure up associations and interpretations, then we cannot make a distinction between the literal use of a word and the figurative use. All words are metaphoric, in a sense, because all words name reality according to its essences. Semantics is not about representing reality, Paul de Man writes, but about translating it (16). Metaphor is no more an act of substitution or comparison than is any act of attaching a word to an aspect of reality in order to represent that reality.

Richards proclaims that we must renounce the idea that words have intrinsic meanings, a concept he refers to as the "Proper Meaning Superstition" (11). Words change meaning according to context, Richards argues. Hence, metaphors resonate with multiple meanings—what I refer to as "harmonics." The meaning that a Santa Fe Institute scientist may intend in using a word like "rules" is inherently no clearer than any of the possible harmonic meanings. The harmonics we hear and those we tune out depend largely on the sentences in which we encounter the words, just as in music where the impression created by a note depends largely on how it sounds amid the overall composition. For metaphor theorist Richard Johnson-Sheehan, it is the audience that determines the meaning of a metaphor according to a context that entails the audience's prior beliefs and the rhetorical situation in which audience members hear or read the metaphor (60–61). This is the argument that Wittgenstein would make in his later work when he claimed that language is a game to which one must know the rules, that is to say, the context.

Let us pause to see where we are in the pursuit of insights into the question, "What is a metaphor?" It seems clear that we can locate a definition of metaphor somewhere along a continuum between two poles: One pole holds metaphor to be a non-constructive, communicative device that delivers knowledge of facts, which exist independent of human awareness. Metaphor from this perspective can aid in understanding reality, but we must assume that literal things exist to be understood and represented metaphorically. The second pole holds metaphor to be a constructive, linguistic device for producing knowledge, which, therefore, must assume that knowledge to some extent is a product of human discourse. This view is associated with the latter Wittgenstein, who held that "human language in a sense *creates* reality" (Phillips 30, italics original). A fact in this light would be the product of the human mind, a description of the attributes of reality, how it appears to us. Reality could never be observed neutrally; as observers we would always help to shape it.

Returning to music theory and science, we could interpret Wittgenstein to say that science scores interpretations of reality against which every new theory resonates. Each metaphor used in science, therefore, is perceived as having some degree of sonority with respect to the larger "score." These scientific scores are similar to Kuhn's paradigms. Scientists continually revise their scores and write new ones. No scientific metaphor stands alone outside of a larger theory, or score, just as no musical note has significance by itself. Metaphor is ostensive in the sense that Boyd argued because it creates meaning by striking our senses.

Hence, the argument about whether metaphors are literal or figurative loses relevance in this musical model. All metaphors have various meanings; one is no more correct or literal than another, any more than one harmonic is more correct or literal than another. Likewise, one metaphor is no more figurative than another. The meanings and images of metaphoric words are simply tones within the larger discourse environment. Some metaphors certainly would be more powerful aids to knowledge-construction in science than others. "Rules" in a social or biological system may be a richer metaphor than the sun as a "furnace" in astronomical science. Still, both constitute theory because both have multiple harmonics that must be accommodated in context of larger paradigms.

METAPHORS IN MOTION: HOW THEY WORK

If we are freed of the "superstition" that words have literal meanings, we are able to open up the realm of metaphor studies and allow metaphors to move among the words and the contexts in which they are used. For theorists in the constructivist camp, metaphoric statements involve what Richards refers to as the "interanimation" between words (69), or what Black refers to as an "interaction" ("More" 27). In both cases, the meaning of a statement is greater—however we might define "greater"—than the sum of its individual lexical components. Richards conveys this notion of a greater sum when he writes that, "a word is always a cooperative member of an organism" (69). Metaphors, in this sense, are vital only when they remain in constant motion; this is an insight that emerges from the etymology of the word. "Metaphor" shares a root word with the Greek terms *epiphora* and *phora*, which suggest the "transference" or "locomotion" of meaning across terms (Lanham 100, Peters 157). Other sources suggest that the second part of the word relates to the Greek term, *pherein*, which means, "to bear" ("Metaphor" *Webster's*). Such movement suggests that truth can never be fixed in language, but must be constantly renegotiated (constructed) as new metaphors appear.

It is significant that the Greek philosophy upon which any theory of metaphor derives was repeatedly drawn to the relationship between things (or essences) and change (or motion). The Platonic distinction between "being" and "becoming" is related to knowledge, *episteme*. For Plato, knowledge could be seen either as the constant movement of the soul toward truth or—considering an alternative etymology of *episteme*—as the cessation of such movement, the arrival at truth (*Cratylus* 437a,b). The Greek study of harmonics that we have been exploring analyzed music as motion in time. The more empirical theorists like Aristoxenus described music as motion against a background of context, where context would be the musical interval. Likewise, for Nicomachus, the neo-Pythagorean, the planets produced musical pitches because they were in "deprived of stasis" (Mathiesen 396). Consistent with this body of theory, we could see metaphors constituting scientific knowledge by oscillating with harmonic movement against the background of a particular epistemology or scientific paradigm.

In proclaiming metaphor to be an act of motion, Richards and Black implicitly are rephrasing the static theoretical question, "What is a metaphor?" in a more dynamic form, "How does metaphor work?" They do so by employing metaphoric language to describe how metaphoric statements interanimate or interact. In examining how metaphoric statements work, they are also examining how they aid in cognition. Richards labels the parts of a metaphoric statement "tenor" (also referred to in the literature as "topic") and "vehicle," where the former is the "plain meaning" and the latter, the figurative (100). Looking at the statement, "The sun today is a furnace," using Richard's model, the word "furnace" becomes a vehicle that carries new associations to the plain word, "sun." Black develops a similar model, where the new associations act like a work of art that changes "focus" when painted metaphorically, but which is always surrounded by the so-called literal term—the "frame" ("More" 28). So our plain image of the sun upon closer focus frames a fresh image of a raging hot furnace. For Black and other interactionists, the metaphoric process moves in two directions, which simply means that the tenor and vehicle affect each other. So, in the example we have been using, we see the sun in the sky as a furnace, but at the same time, our mental image of a furnace also becomes more like our image of the sun.

Perelman and Olbrechts-Tyteca offer similar views of metaphor in action by showing metaphor to contain associations that were at one time expansive, but have been fused into a terse, powerful statement pregnant with cognitive import. The two rhetoricians define the

collapsed metaphor in terms of the larger analogy, where the relationships between two terms are said to be similar to the relationship
between two other terms. We could extend our example: "As the furnace forges the copper ingot, so the torrid sun forged the earth."
Perelman and Olbrechts-Tyteca would call the image of a furnace producing copper the "phoros," which recalls the Greek term for transference or locomotion. Here it carries new meaning to the relationship
between the sun and the earth. In the most powerful metaphors, the
two most important parts of the analogy are joined, as in, "The sky's
furnace forges the earth." Perelman and Olbrechts-Tyteca conclude
that, "Metaphorical fusion does not involve closer relations between
phoros and theme than exist in simple analogy, but its effect is to consecrate the relation between them" (401).

"Consecrating relation" and "transferring knowledge" are potent
images with which to gain awareness of how metaphor serves to construct meaning. All of the constructivist theorists have written insightful phrases full of their own metaphors in an attempt to capture how
metaphor does this. Black argues that metaphors project a series of
implications from one concept to another ("More" 28). Ricoeur
writes, ". . . it is from metaphor that we can best get ahold of something fresh" and that metaphor is "the power of making things visible,
alive, actual . . . " (33–35). Metaphoric statements for literary critic
Monroe Beardsley have a progenitive power to bring to the surface
hidden connotations of meaning that may "wait, so to speak, lurking
in the nature of things, for actualization—wait to be captured by the
word" (quoted in Ricoeur 97).

Assuming that metaphors serve to highlight relations among different objects, actions, and ideas, some metaphor scholars have pro-
ceeded to classify metaphors according to type. This suggests that
individual metaphoric statements all fall under a taxonomy of
metaphor. George Lakoff and Mark Johnson's study, *Metaphors We
Live By*, provides such classification. Metaphors function by calling
attention to spatial orientation (up versus down, as in, "Things are
looking up."); ontological status (in versus out, as in, "He is in love.");
causation (where creation is akin to birth, as in, "The University of
Chicago was the birthplace of the nuclear age."), and so forth.

Philosopher Stephen C. Pepper narrows the focus of Lakoff and
Johnson's argument, making a Platonic/Kantian claim that all reasonable perspectives from which people perceive the world have at their
core just a few "root metaphors." Pepper's use of the word "metaphor"

is peculiar, implying more the concept of paradigm than of metaphor. Pepper argues throughout *World Hypotheses* that four philosophical perspectives are adequate for understanding the way that contemplative people view the world. They are "formism," founded by the metaphor of "similarity," which holds that because the world is full of things that seem to be alike, reality is a manifestation of Platonic forms that organize those things (151); "mechanism," whose root metaphor is the "machine," suggesting that reality is a mechanical structure of connected parts (186); "contextualism," a type of pragmatism with "historic event" as the root metaphor (232); and "organicism," which holds that terms from biology, "organism" and "integration," underlie a reality in which the parts interact to create a system that transcends the fragmented appearance of that reality (280). The last worldview seems strikingly close to that found at the Santa Fe Institute.

Metaphors such as those cited by Lakoff and Johnson, or even the metaphoric categories of Pepper, assist in cognition by drawing attention to similarities between things or phenomena. But, in calling attention to similarities, metaphors implicitly also call attention to differences. For example, the development of nuclear science is similar to birth in that new theories of physics emerged like a baby created by the union of ideas from various scientists at a university. But the development of such a science is different from birth because development of the new science took longer than nine months and had more than two parents. Perelman and Olbrechts-Tyteca developed the taxonomy of "dissociations" to shown how cognition is aided by noticing these differences. Common themes of dissociation, for example, involve the difference between "appearance and reality," "means and ends," "multiplicity and unity," and so forth. Using this technique, we could unpack metaphorical similarities to reveal their deception. For example, we could argue that while the sun may appear to be a furnace, it is not because it is not a man-made structure. So while metaphor aids in cognition by drawing attention to the ways in which one part of the world is associated with another, it also does so by showing how these parts of the world are different.

Various empirical studies have devised human-subject experiments to examine how metaphor works to assist in knowledge cognition. It is clear from a cursory review of the research literature that metaphor deeply affects the way we think and view the world. Yet, the results show that metaphor is a complex part of discourse, and that no one theory adequately covers all instances of it in action. Ortony and fellow researchers, for example, show in a 1978 study that metaphor theory cannot be reduced simply to a study of how words are substituted for one another or compared with one another; metaphor is not just a lexical phenomena, but one which is highly dependent upon context and discourse situation ("Some Psycholinguistic Aspects"). Subjects in the Ortony study understood metaphors much better when the metaphors were set up by several sentences of explanation. For example, when subjects heard the metaphoric statement, "The fabric had begun to fray," in reference to a troubled marriage, they understood it when it followed a full paragraph discussing a couple's marital troubles. The subjects had a harder time following the frayed fabric metaphor, however, when it followed only the single sentence, "Lucy and Phil needed a marriage counselor" (79). This research from Ortony and his colleagues shows that metaphoric statements are valuable aids to cognition when they are carefully used and supported by a more literal explanation. In this sense, perhaps metaphor functions as a catalyst that transforms the ordinary discourse surrounding it into a powerful summarizing image.

Robert Verbrugge's research explores this transformational component of metaphor, and suggests that metaphor functions by stimulating the human imagination. "At the outset, these results support the claim that *fanciful* experiences are common when comprehending metaphor" (113, italics original). Verbrugge extends Black's interaction model to claim that tenor and vehicle do not merely interact, but that the tenor is transformed into the vehicle. In his study, published in 1980, Verbrugge and colleagues gave 25 people each a list of ten sentences with metaphors and similes and asked them to write down comments about each. The subjects described images that came to mind after reading statements like, "Skyscrapers are giraffes." A typical response was, "I saw large, tall, windowed buildings which became very skinny and developed spots" (110). This response suggested that the reader mentally converted the literal image of an inanimate skyscraper into the figurative image of a living animal. Such conversion happened more often with metaphor than with simile.

These results and others show that humans comprehend their world through metaphor by developing mental images that extend beyond the literal meaning of individual words. Richard Harris, Mary Anne Lahey, and Faith Marsalek in a study published in 1980 examined the way in which 78 undergraduate students used images to recall groups of sentences. Each group had a metaphoric statement ("The ivy cuddled up to the window."); a dead metaphor statement, where the metaphor was said to be so common that it was no longer recognized as fresh ("The ivy crept up to the window."); and a non-metaphoric statement ("The ivy grew up to the window.") (167). The subjects were told to commit the sentences to memory. Results from this experiment suggested that people use imagery more when thinking about metaphoric statements that are fresh than when thinking about other kinds of statements, although the results did not reach statistical significance (169–170).

Clearly, humans use their imaginations when responding to metaphors, and this process of creating mental images helps them to produce and understand knowledge. But the way in which individuals deal with metaphoric images is subjective and depends upon on what could be seen as intangible factors. For example, the Verbrugge study shows that people seem to have an innate preference for what part of a metaphoric statement should be the tenor and what part should be the vehicle. The statement "Skyscrapers are giraffes," for instance, was preferred to the inverse, "Giraffes are skyscrapers." A study published in 1980 by Kathleen Connor and Nathan Kogan based on research with 136 students found that they also exhibited directionality preferences. When given a pair of words like "ancient tree" and "grandfather," for instance, the subjects clearly preferred metaphors in which the grandfather was the tenor and the tree was the vehicle. The researchers concluded that people prefer metaphors where human beings are the tenor, e.g., where grandfather is likened to an ancient tree rather than the old tree likened to grandfather (300).

Michael Johnson and Robert Malgady in various studies in the late 1970s explored what makes people like a particular metaphor. Subjects were given pairs of words like "snow" and "confetti" and asked various questions about the pair. Subjects rated those pairs to be "good" when they found that they had many attributes in common and when the relationship between the two was easily interpretable (272). Much of this research shows that people have preferences when dealing with metaphors and that not all metaphors are created equal. Of particular note from the two directionality studies discussed earlier, for example, is that in one case the inanimate object (skyscraper) was preferred as the topic, while in the other case the living being (grandfather) was preferred. Clearly the effectiveness of a metaphor cannot be determined by fixed rules; metaphors are useful tools of cognition in part because of the way they strike the ear in association with the belief systems of an audience. Some apparently just sound better than others.

This discussion of metaphor theory has covered a lot of ground in pursuit of answers to the questions, "What is a metaphor?" and "How does metaphor work?" Some insights have emerged that would seem to be a solid foundation for a study related to rhetoric and metaphor at the Santa Fe Institute. Certainly we can agree that metaphor functions across a range. It helps to deliver or make accessible the knowledge necessary for understanding and, more important for this study, it produces knowledge. Metaphor functions by allowing two concepts to interact in the human mind, creating knowledge that transcends that which is held in the individual concepts. When we hear that, "The sun today is a raging furnace" we know more about the environment outside than if we were merely told that the weather is hot. Something happens in the mind that fuses the images of the sun and the furnace together to offer an awareness of conditions outside that would not be available through non-metaphoric statements. Yet, as we have seen, it is difficult to measure and explain what happens in the mind as it creates or responds to metaphor. Human cognition remains a mysterious process, but it is clear that metaphor has a deep and integrative role in that mystery.

METAPHOR AND SCIENCE

Having explored the theory of metaphor and how it works, it is now time to return our attention to the role of metaphor in science. We have seen arguments that metaphor in science can function at various places along a continuum—as a tool to build knowledge or, at the other end, as a figurative device that may have some epistemological value, but which primarily serves to help deliver knowledge. Yet, an earlier conclusion that was starting to unfold should become more apparent in this section: Nearly all metaphors have some role in producing or fine-tuning knowledge.

Science has never been without metaphor and related figures of thought and speech. Examples are everywhere; where there has been discovery and invention, there has been the language of metaphor to align those new insights with old ones. One example of scientific metaphor serving primarily as a figurative device used to aid in understanding of new knowledge is found in the work of Nicholaus

Copernicus. In the early sixteenth century, Copernicus revolutionized our image of the solar system by showing it to be heliocentric. "For who would place this lamp of a very beautiful temple in another or better place than this wherefrom it can illuminate everything at the same time?" he asked, rhetorically, in *On the Revolution of Celestial Spheres*. "As a matter of fact, not unhappily do some call it a lantern; others, the mind and still others, the pilot of the world. . . . And so the sun, as if resting on a kingly throne, governs the family of stars which wheel around" (71). It would be hard to imagine a passage of science that had more metaphors. We can picture Copernicus, unable to contain his awe for the solar system, feeling compelled to heap metaphor upon metaphor in an effort to honor, understand, and explain it. Although these metaphors seem overly poetic, they could be seen as having a small role in helping to build knowledge. They emphasize the dominance of the sun in the solar system—a claim that is obvious now, but was once revolutionary.

As the previous discussions have shown, many rhetoricians of science and scientists themselves would acknowledge that metaphor does more than simply illustrate poetically the literal world of science. The history of science is replete with examples of metaphorical insights that led to breakthroughs in knowledge when seemingly more conventional methods of theorizing reached an impasse. Organic chemistry ascended in the mid nineteenth century when experimental chemists postulated that carbon atoms could link together by means of "hooks," which they called valence bonds (Asimov 21). Russian chemist Demitri Mendeleev in 1869 gave order to the panoply of chemical elements by arranging them in a periodic table; he likened properties of the elements to musical harmonics, where sound qualities repeated periodically (quoted in Hoffman 411). It is interesting to compare Mendeleev's use of the musical term "harmonics" to show repeated patterns in chemistry with my use of the same concept in this book to show submerged connotations within a metaphoric word.

This anecdotal evidence of metaphoric theorizing reveals metaphor to be an aid to knowledge construction throughout science, but these examples do not resolve the philosophical questions related to its function. Although we may be convinced that the metaphoric process aids in knowledge production, we have no grounds to make the leap that just because a particular metaphor is present in science means that particular metaphor was necessary to create the knowledge it conveys. Perhaps Mendeleev first recognized that certain qualities of chemical elements repeat themselves, and then he metaphorically conceived of that repetition in terms of musical harmonics. If so, the metaphoric act would be that of naming something already conceived non-metaphorically. Metaphor could be seen here as a device to help Mendeleev understand the knowledge that his research had led him to produce non-metaphorically.

In the context of scientific investigations, Boyd shows that the question of which came first, the metaphor or the knowledge, is not as silly as it sounds. In some cases, scientists develop metaphors because they have no other means of conceiving of an idea, Boyd writes (360). These are "theory constitutive," a term that is similar to the idea of a constructivist metaphor, but perhaps more sweeping—or "inductive open ended." Here a single metaphor can contain the germ of an entire theory and can induce new aspects of that theory (370). The example Boyd cites is that of computer systems said to behave like the human mind, an example we will look at in some detail in the analyzing Santa Fe Institute metaphors. The metaphor is powerful because it invokes associations between mind and machine even though "we still do not know in exactly what respects human cognition resembles machine computation" (370). The metaphor of the computer as a mind is scientifically imprecise, and while Boyd accepts such imprecision, he adds that for any theory to have heuristic value, it must be "in some important respects approximately true" (401). In Boyd's analysis, scientific truth is similar to the positivist notion of "laws," whereby natural phenomena are classified according to the laws that cause them to be. So, presumably, for the computer metaphor to be valid, a computer must cause information to be processed in a way that approximates the way in which the mind processes data so as to become knowledge.

Metaphor theorist Zenon Pylyshyn would go further to argue that imprecision is not only acceptable in theory-generating metaphors, but also essential because these metaphors are operating in realms of science not fully charted. Pylyshyn quotes Boyd: "Use of metaphor is intended to introduce terminology for features of the world whose existence seems probable, but many of whose fundamental properties have yet to be discovered" (429). Hoffman shows that the best metaphors in science are those that spawn theoretical ponderings over many years, such as the metaphor of light as a "wave" or a "particle." "In science," he writes, "one can latch upon a metaphor or intuitively appealing vision (e.g., waves) and ride the vision for years, or generations, trying to unpack its implications (e.g., Bohr equations, wave particle dualities, etc.)" (415).

Yet, Boyd holds other metaphors to be "exegetical" or "literary," which means that they follow theory and explain it without being essential to the genesis of the theory. These non-constructivist metaphors are "conceptually open ended" because they help create new ways of envisioning an existing theory. He gives the example of early modern physicists, including Niels Bohr, who worked with models of the atom resembling the solar system. For Boyd, the metaphors embedded in this model are of a limited teaching value, but do not conjure up new aspects of the theory. To proceed further in explaining atoms a scientist must go well beyond the solar system imagery in order to capture the quirky erratic behavior of subatomic particles. As Boyd writes, "The function of a literary metaphor is not typically to send the informed reader out on a research project" (363). But in the theory-constitutive computer metaphor example, by contrast, researchers are compelled to explore new ways in which a computer and human brain behave similarly and differently.

Pylyshyn explores Boyd's distinction between the theory-creating and literary teaching functions of metaphor in science by invoking Jean Piaget's psychological theories of intellectual development. A literary metaphor requires a process of "assimilation," whereby the metaphor makes new ideas accessible to the individual's intellectual system, or "schemata," Pylyshyn explains. A theory-constitutive metaphor requires "accommodation," in which the schemata change in response to the new information (421). Kuhn, however, takes some issue with this refined distinction. While acknowledging that some scientific metaphors only go part way in explaining a phenomenon, Kuhn argues that in that case, scientists develop new metaphors to fill in the blanks. So when the solar system model of the atom falls short, scientists move to metaphors borrowed from mechanics, where electrons and nuclei might be seen as charged billiard balls. The scientists then refine the theory and the model to see where similarities apply, and where they do not ("Metaphor in Science" 415). Even an imperfect metaphor would be theory generating for Kuhn because it prompts researchers to question which aspects of the metaphor are appropriate and which are not. This method of developing an image and then showing where reality matches the image and where it does not recalls the Olbrechts-Tyteca notion of argument by association and dissociation. For Hoffman, the chance that a metaphor may be wrong in certain aspects is a property of metaphoric theorizing that scientists operating out of the positivist tradition should appreciate, for it allows aspects of the theory to be easily falsified (402).

As we have seen so far in this brief look at metaphor in science, the general question of whether metaphor is a device for producing knowledge or merely one for delivering it has been recast using the terms "theory constitutive" versus "literary." A neural psychological approach to the question probes the actual process of thinking to ask whether metaphor causes our brain schemata to change to accommodate new insights, or whether metaphor merely allows us to assimilate new understandings through existing schemata. It seems unlikely that such a question involving brain physiology could ever be answered. This question of which came first, the metaphor or the knowledge, is not easily resolved into a general rule that would apply in all cases. Perhaps we are left to conclude that metaphor in some cases produces scientific knowledge by changing the wiring in a scientist's brain to allow him to see the world differently, and in other cases it merely makes that knowledge more palatable to the scientist's existing ways of thinking. Or, taking the Kuhnian position, we could argue that the distinction is irrelevant; metaphor always affects cognition regardless of the specifics of how it operates. In the same way, every note and its harmonics affect how we respond to a piece of music. We must "accommodate" each note that a composer adds. The evidence forthcoming from the Santa Fe Institute interviews would seem to support such a Kuhnian argument in science writing.

FEAR OF PERSUASION AND METAPHORIC HARMONICS

Yet, even as scientists have depended upon metaphor to produce knowledge, perhaps they have done so in some cases because they were stuck and needed an extraordinary aid to cognition. Arguably, metaphor could be seen as a "trick," the way identities often are used in mathematical equations to help the mathematician replace one algebraic phrase with an equivalent one in order to move beyond an impasse. While Hoffman does not say so, he suggests that scientists often do not respect metaphor because of its trickiness. Most scientists would acknowledge that metaphor works, but they then might invoke the commonplace argument that theories based upon language are not as "rigorous" or "robust" as are other theories. Hoffman writes, "The general rule some scientists and philosophers seem to follow, today as in the past, is this: If you don't like a theory (or theorist) and if the theory is metaphorical or contains metaphor, you can usually get away with criticizing the theory by saying, 'It's only a metaphor'" (402). Much of twentieth century psychological science, for instance, was premised on the notion that "theories were supposed to avoid all the vulgarities, ambiguities, mentalisms, and vagueness of ordinary language" (George et al. cited in Hoffman 394).

No doubt some of the suspicion of metaphor in science is linked to the suspicion of rhetoric in general, and the persuasive aspect of rhetoric in particular. Metaphors, as one type of rhetorical trope, cannot escape the taint of persuasive discourse. Ricoeur's prescient analysis of metaphor shows that the persuasive component of metaphor is integral, and that any metaphoric thinking that attempts to avoid persuasion loses its potency. He is writing about philosophy in the following passage, but the same argument applies to science: "Philosophical discourse is itself just one discourse among others, and its claim to truth excludes it from the sphere of power. Thus, if it uses just the means that are properly its own, philosophy cannot break the ties between discourse and power" (11). It follows that a metaphor that provokes new scientific thinking must be imbued with power, which means that it must be persuasive. The metaphor of the human mind as a computer does not neutrally represent reality; it *persuades* us that the association is real. It follows, then, that just as all metaphors would seem to have some role in producing knowledge, all metaphors also would have a role in delivering that knowledge persuasively.

Ricoeur and others show, however, that metaphors can be persuasive in ways that may seem dangerous because, by definition, metaphors are only vital when they remain in constant motion. This argument follows from the etymology of the word "metaphor," which we have seen suggests the transference of meaning across terms. Ricoeur notes that meaning is unstable, highly dependent upon context (127). A scientist who employs a metaphor does so, knowingly or not, at the risk of losing control of that metaphor—of having it "move" in unintended ways. Metaphor and other rhetorical devices of transporting knowledge are linguistically corrupt, as Poststructuralist philosopher Jacques Derrida argues throughout his treatise on metaphor, "White Mythology," because the metaphoric process invariably borrows so much from one sign that it distorts another. Such distortion does not cause concern for Derrida; it is intrinsic to his concept of language. In Poststructuralist theory, a signifier is never exclusively attached to the signified, or to whatever object it points. The signifier constantly conjures up new meanings. Derrida in his writing refers to this as "freeplay," a concept that is identical to my concept of metaphor harmonics.

Nonetheless, it is easy to see how a scientist could find these harmonics unsettling. The scientist using a metaphor has no way of knowing whether the harmonics will elicit new insights, as in the computer-as-brain metaphor, whether the harmonics will be harmless, or whether they will distort meaning (by offering too many "tones" that may distract from the representation that the scientist was trying to reach). Hoffman shows, for example, how one particular anthropo-morphic metaphor could spawn harmless unplanned associations in physics, but misleading ones in biology. We can speak of an elemen-tary particle in physics "feeling a force" without risking the false impression that particles actually feel, Hoffman writes. But the same metaphor applied to cells "feeling the effect of toxins" could give the impression that cells are of a higher life order than they are, and hence "feel" in the way a whole living organism might feel (399). Some cells and bacteria, of course, do respond to chemical stimuli in an involun-tary process known to biologists as "chemotaxis." But to say that they "feel" the stimuli suggests a sensory process that involves some kind of cognition beyond mere chemical response.

Young examines the implications that oscillated from Darwin's metaphor of natural selection—implications that remain powerful today. Darwin described a process whereby an organism that is fit is more successful at reproducing itself than is a less fit organism, which means that the fitness is passed on, while less fit versions of the organ-ism are not. He was referring to a purely natural process, yet the selec-tion metaphor implies a kind of anthropomorphic agency, or sense of purpose, as if some powerful force entity (God) did the selecting. Philosophers of science like Dennett have argued that metaphors of this kind have masked the stark realities of what he refers to as "Darwin's dangerous idea." For Dennett and other neo-Darwinists, such metaphoric overtones may give people false comfort in old human-centered religious belief systems, rather than prompting them to develop ethical systems that celebrate the diversity of all kinds of evolved life.

Metaphors change science not only by creating unintended har-monics, but also by sometimes reducing a difficult and uncertain con-cept to images that are too easy to understand or paraphrase without subtlety. For example, economist Deirdre McCloskey has argued convincingly in *The Rhetoric of Economics* and other texts that as a social science, economics relies heavily upon metaphor, narrative, and other literary tropes. The best-known example is Adam Smith's metaphor of the "invisible hand" to describe events in the marketplace. Yet, economist Robert Solow argues that Smith's metaphor has been subject to reductionist thinking because it implies that free markets are always good, even though Smith and every other economist since has recognized that markets only work well when information and access to those markets are available to all members of society—a heroic assumption in the real world. So the metaphor turns a positive description of economic science under rarefied conditions into a normative prescription for laissez-faire policies because the image of an "invisible hand" is an image of such avuncular goodness.

Solow offers other examples of how he sees complex theories turn to "mush" when appropriated by the language of the lay public. He does not deny the presence of metaphors and stories in economics, but suggests that mathematical metaphors are more productive for economists than are literary metaphors (34). A mathematical assertion about complex numbers lying on a plane, for example, is metaphorical, but open to research in the way that a literary metaphor, "you are my sunshine," is not, Solow says. Statistical evidence is more valuable than anecdote in economics, Solow argues, because bias is easier to detect in statistics than it is in anecdote.

Ironically, such arguments from Solow and others imply that many economists and other scientists fear metaphors and other rhetorical devices primarily when they are used as tools for delivering ideas to the lay public, rather than when the metaphors are kept relatively hidden inside the confines of science for the purpose of aiding in knowledge production. The fear is that when metaphors are used to convey scientific knowledge, the tropes that solidify in the public canon often are more powerful than the discipline-specific concepts that gave birth to them. For example, common terms like "inflation," "government spending" and "growth" pulsate with multiple overtones, which are not present in the equations that cast these concepts into relationships with the larger economic world. "Government" in the Depression-era of economist John Maynard Keynes, for example, was what rhetorician Richard Weaver would call "a God term," representing salvation from the despair of a dark and terrible economy (Foss et al. "Richard Weaver" 73). In the era of rational-expectations economics, or millennialist conspiracy theories, "government" became a "Devil term." In the wake of the 2001 terrorist attacks, "government" may once again be venerated as we rally around a protective authority. Yet, the mathematical symbol for government spending in the popular general equilibrium model of national income has not changed. What has changed are the rhetorical values assigned to that symbol.

Bernard Cohen, a historian, has studied the way that new knowledge from Newtonian physics percolated into the eighteenth century political philosophy that formed the foundation of the U. S. constitution. Cohen's research suggests that government founders— men like Thomas Jefferson, John Adams, and James Madison—relied on often-faulty interpretations of scientific concepts and metaphors in arguing for or against a political balance of power. Cohen's point is not that the U.S. constitution would have been crafted better if the scientific terminology, such as "equal and opposite reaction," "laws of nature," or "equilibrium," had been more accurately cited. Instead, he shows that highly specialized scientific terminology has a way of infiltrating the popular lexicon in ways scientists cannot predict.

We see the difficulty of controlling metaphors and how they will be heard. The lesson from this section is not that metaphors potentially are so misleading as to be of limited value in science, although it is clear that some scientists might lean toward such dismissal. Rather, scientists must accept that metaphors are powerful cognitive devices because they are some of the most evocative terms in language, which out of its human origins is evocative and unsettled. Scientists debate word usage, as so they should, for only in doing so are they able to hear the full implications of their ideas.

RHETORICAL ALTERNATIVES TO METAPHOR IN SCIENCE

It is the potency of metaphor that has led some philosophers of science to argue that similes, analogies, or models are preferable as a means of coaxing forth new theories. The terms all share a related function, which is to draw attention to resemblances. The word "simile" is a Latin term meaning "like," comparable to the Greek term, *eicon*, which suggests an "image" (Lanham 140, 89). Similes overtly state that something *is like* something else, and therefore, call attention to the act of comparing in a way that metaphors do not. Arguably, a simile is less dangerous than a metaphor because it acknowledges the comparison and invites rebuttal. (Recall, we have seen this argument at the Santa Fe Institute). If a scientist says that an atom is like the solar system, he or she is stating the comparison overtly and inviting someone to point out discrepancies—ways in which an atom is not like the solar system. But the scientist who employs solar system metaphors for atomic particles, perhaps describing the "orbit" of an electron, is invoking a similarity without calling attention to it. Doing so suggests a certainty to the comparison that may not be justified.

In some instances, then, simile may be more appropriate, especially when a scientist is groping for comparisons that are new and as likely to be inaccurate as not. Yet, Ricoeur argues that because simile lacks the power of metaphor, it also lacks the potential to shape cognition. In his words, simile "dissipates that dynamism of comparison by including the comparative term" (26). Ricoeur's analysis borrows heavily from Aristotle, who argues that simile and analogy are forms of metaphor. For Aristotle, the simile is "less pleasant, as it is more drawn out, and it does not say that this is that, and so the mind does not think out the resemblance either" (*The Art of Rhetoric* 235).

Analogy, as noted earlier, draws extended metaphorical comparisons by suggesting that two things are alike in the same way that two other things are alike. The Greek origins of the term suggest "equality of ratios" or "proportion" (Lanham 10). Hence, the pair X and Y relate to each other *as if* they were the pair A and B. Gentner and Jeziorski note that analogy is a selective system of comparison, whereby certain common features are identified and others ignored. In their example, we might agree that a cell takes in resources in order to function as if it were factory taking in the raw material of production. But we would ignore non-related aspects of the analogy, recognizing, for example, that a cell is not made of bricks and steel (448). Once again, the overt lexical marker "as if" shows the analogy for what it is and spells out exactly what is comparable and what is not. Such honest use of language perhaps avoids spawning the unwanted harmonics of metaphor, but it also restricts the comparison's potential to generate unexpected, but useful new theoretical implications.

Analogies when extended in science are known as models. Black argues that models can be seen as exact representations of phenomena, where specific details in the model correspond to specific details of the phenomena. An icon of a Catholic saint, for example, provides a small image of a larger human figure, while a model airplane shows the scaled-down dimensions of the real thing. Scientific models are not icons, but are "analogues" in Black's terminology. So, for example, the model of an atom as the solar system implies an analogy between the nucleus as the sun governing the behavior of electrons as planets. The model of the brain as a computer implies analogies between the various physical components of each. Black writes: "An analogue model is some material object, system, or process designed to reproduce as faithfully as possible in some new medium the structure or web of relationships in an original" (*Models and Metaphors* 222, italics original). Models do not provide proofs, but hypotheses of how things work, Black says. They are "speculative instruments," he adds, quoting Richards. For Black, a model differs from a metaphor in that a model requires prior control of scientific theory, while a metaphor requires only commonplace associations.

Black's distinction between models and metaphors reveals the power of each and the order in which each is found in science. First come the metaphors, which force a scientist to accommodate the various meanings associated with the metaphor. This can be likened to the way a musician must accommodate the various harmonic tones whenever calling upon an instrument to add its notes to a piece of music. Successful metaphors in science—those that are sonorous and not spurious or dissonant—form the primary basis of comparison. From these, more elaborate analogical connections can be extended. For example, once scientists accept that the brain is a computer, they can then draw specific appropriate connections between memory functions, calculating functions and so forth, while rejecting inappropriate connections. But without the initial metaphor, the model has no power. The importance of metaphor as a tool for theory building in this way will become more obvious as we look in detail at how metaphor is used at the Santa Fe Institute.

4

Metaphors and Mathematics: A Shared Tradition of Constituting Knowledge in Science

At the beginning of this book, I posed the question, "What can the rhetorical challenges confronting members of the Santa Fe Institute tell us about the role of rhetoric—specifically metaphor—in science?" Perhaps a short answer can be found in the comments of Erica Jen, then vice president for academic affairs, who was quoted earlier. In consoling me for a failed attempt to capture in words the concepts of non-linearity and discrete time, Jen intimated that some aspects of scientific reality are nearly impossible to express. Jen was restating one of the fundamental problems of the epistemology of science: How can we claim to know reality? This is a philosophical problem not exclusively restricted to language. The truth of scientific knowledge depends on, among other things, the limits of our senses to perceive the world around us and the preconceptions we may have about reality that affect *a priori* the reality that we choose to see. The limitation of our senses is a physiological problem, while the preconception problem is, to some degree, a psychological or sociological matter.

Yet, once we observe something—however imperfect our observations may be—we enter the linguistic realm in trying to describe it and represent it. Word meaning can be maddeningly elusive, as we saw earlier when a scientific meeting became tangled in the effort to define the word "rules." This chapter explores the insights of members of the Institute who recognize the challenges of creating and representing knowledge in language. I present the results and analysis from seventeen interviews that I conducted with SFI members during the summer and fall of 1999. I also draw on texts written by these and other Institute members, and from the body of literature dealing with the philosophy and rhetoric of science. These interview results can be read as a dialogue among the Institute scientists and other living and dead scientists, philosophers, and rhetoricians of science, whose ideas form the canon of scientific literature. To help the reader avoid confusion when moving among interview comments, quotations from the literature, and my own analysis, I have occasionally bracketed the initials (SF) after a Santa Fe Institute member's name when I felt it necessary to emphasize that the comments came from an interview. The dialogue here quickly reveals that all of the Institute members interviewed recognize the importance of metaphor and other rhetorical devices in science, but some would lean towards Jen's appraisal: sometimes even the best metaphors do not get close enough to the essence of reality.

Most of the categories of discussion topics relate to a central problem, the role of language in creating and representing knowledge at the Santa Fe Institute. The discussion concerned the value of metaphor as a tool for creating knowledge and for presenting it (or re-presenting it), while also exploring some of the properties of metaphor—namely unintended connotations and implications, which I refer to as "harmonics." This discussion about metaphor often looked at specific metaphors at the SFI and considered whether they would be theory-constitutive or literary ornaments. Often I found scientists referring to various terms they used as metaphors, although it was not always clear that those terms were metaphoric. Particularly in the case of the word "complexity," it seemed that scientists used the term "metaphor" to refer to any evocative language whose meaning was elusive.

When the reader finishes this chapter, various insights should be clear about the use of metaphor and the problems of representing knowledge at the Santa Fe Institute. Broadly stated, it should become immediately obvious that metaphor has a theory-constituting role at the Institute. Metaphors do so because they are dynamic; they stimulate interactions among ideas. Here I am espousing a Poststructuralist argument, which is that metaphoric language, even when written, is in flux. There is no "proper meaning" to metaphors, Plato's idealism notwithstanding. Meaning is subject to the interpretations of an audience that responds subjectively to how words strike the senses in the context of other words. Yet, metaphoric language also has Platonic properties in that by inviting associations, metaphors create an implied dialogue among scientists struggling to find the illusive "right word." This is the paradox: No right word exists, but the search for it in science is what gradually and imperfectly constitutes knowledge. Specifics of the theory I advance through these interviews build from the following points:

- Metaphor constitutes theory by ringing forth with various signals, various meanings. A metaphor like "fitness land-scape," for instance, transfers impressions of a geographical landscape to a biological theory of evolution. Even when a metaphor's meaning is vague, such as in the term "edge of chaos," it can be useful for theory building by inspiring scientists to think broadly. To use the language of Richards, metaphor brings many "vehicles" to bear on the "tenor." A metaphor in science brings with it one strong association, often intended by the user, and many other associations—harmonics that may not have been expected or intended.

- These harmonics usually are valuable because they force the theorist using a metaphor to confront implications of a theory. Sometimes in science those implications may contradict the main premises of a theory, or render it paradoxical.

- A scientist forced to reconcile the paradoxes of her theory is performing the cognitive act of "accommodating," as we encountered earlier in the literature of metaphor theory. Accommodation occurs when an individual's schemata for understanding the world change in response to the new information contained in the harmonics.

- Metaphor harmonics often lead Santa Fe Institute scientists to confront philosophical questions that have perplexed scientists and philosophers for millennia. Some of these questions are epistemological, asking what constitutes knowledge. Others are metaphysical, asking if anything intercedes between material substance and events in the world. These harmonics conjure up questions about the nature of reality and whether that nature can be represented accurately and still be understood. The harmonics also invoke questions about intentionality, or will; questions about the definition of life, individuality, and autonomy; questions about the nature of time and causality; and questions about whether reality at all levels behaves according to similar rules.

- Not all metaphor harmonics provide such valuable stimuli for new ideas. As in music, some harmonics add to the sonority of a piece, but too many would affect its overall timbre, perhaps creating unintended or even unpleasant sounds. Metaphor harmonics can be particularly noisy and disruptive for Santa Fe Institute scientists when they emerge from words used by the Institute that also are common in the non-scientific community, or when those harmonics suggest simple applications for abstract theories. Harmonics can be particularly dangerous when they deface the essence of what it means to be human by drawing inexact parallels between human agency and the laws of physical action.

- Invariably, any discussion about making theory in science leads to a philosophical debate about the relationship between verbal knowledge, including that borne out of metaphor, and mathematical knowledge. SFI scientists seem to embrace metaphor freely, but many hold that verbal language cannot always capture their ideas as well as the formal language of mathematics. The tension between these two syntactic systems persists in this complexity science even though it is new and highly philosophical. Some scientists at the Institute who see metaphor as useful also see it as a temporary aid to thinking in that it lays the foundation for more rigorous mathematical theories. It should become clear that mathematics is necessary, even indispensable, for developing scientific knowledge. Yet, a crucial caveat for technical writers who aspire to offer assistance to scientists is that mathematic systems of organized thought are not sufficient for ensuring human understanding. This act also requires images and narrative, and the management of word harmonics.

THE METAPHORS OF INFORMATION AS THE NEW MATERIALITY

Ultimately, these discussions at the Institute about the techniques of developing and presenting knowledge evolved into discussions about the knowledge that is emerging and how it relates to the language used to express it. Before engaging the interviews, it is worth digressing to underscore a theme that loosely connects the metaphors we will encounter. They reveal the postmodern science of the SFI to be clearly metaphysical. That is, it is moving from a focus on the observable material reality of classical physics—which once underlay all sciences—to a new focus on interactions, which are not necessarily observable, among those objects.

In the Classical and Newtonian worlds, reality was constructed out of particles described metaphorically as "building blocks," which suggested that a builder (at first God, later the laws of physics and chemistry) moved about dumb atoms and molecules that had no relationship to each other absent the hand of the builder. Various writers have argued in recent years that Newtonian science was distracted from accurately perceiving the nature of reality because scientists focused on the existence of matter (the particles) instead of on the interactions among that matter that led to higher orders of reality. Such interaction involves the exchange of information among objects. To see evidence of this new perspective on information exchange, we need only look at twentieth century particle physics. There we encounter seemingly farfetched theories of how particles behave according to "knowledge" of how other particles are behaving. With this new physics as a backdrop, the reader of this chapter should come to see that for many scientists, information has become the new building blocks of reality—the new materiality.

Science writer George Johnson has described in his writing how nineteenth and twentieth century scientists came to envision reality not so much as the presence of animate and inanimate things, but more fundamentally as the presence of information in a sea of randomness. A breakthrough insight occurred in the late 1940s when Bell Laboratories researcher Claude Shannon determined how radio and telephone signals are conveyed faithfully despite the presence of noise. Shannon's findings revealed that signals filled with information sent by human beings were subject to the same laws of entropy that govern thermodynamic systems (Johnson, *Fire in the Mind* 122). Information, then, is the ordering process that brings temporary meaning to a reality that is constantly decaying into random noise. Even something as real as an automobile battery could be seen as a repository of information, Johnson writes, in that it begins with a temporary and orderly arrangement of positive and negative charges, but gradually decays into randomness (127).

This paradigm that sees information as the new materiality of science finds favor among many current theorists who posit a fundamental shift in human society from the industrial world of things to the post-industrial world of information, which is mediated through the computer. Science journalist Tom Siegfried captures this shift from Newtonian mechanics to information science in the title of his 1999 book, *The Bit and the Pendulum: How the New Physics of Information Is Revolutionizing Science*. Science and technology writer George Gilder, in his 1989 history of computer sciences, *Microcosm*, heralds a new information paradigm:

> The central event of the twentieth century is the overthrow of matter. In technology, economics, and the politics of nations, wealth in the form of physical resources is steadily declining in value and significance. The powers of mind are everywhere ascendant over the brute force of things. (17)

Many of the metaphors and non-metaphoric terms that have currency at the Santa Fe Institute and other research centers reflect this awareness that information not only describes reality, but also *is* reality. Examples of terms that convey this awareness are "complexity," the study of surprising order arising out of simple rules; "emergence," the claim that properties of systems emerge as the parts of the systems interact; and "information contagion," which shows how the spread of information throughout a human economy affects outcomes in that society. Thus, the pattern of current metaphor use reflects a broad development in the scientific construal of the physical world. These metaphors and related terms would seem to simultaneously construct and reflect that understanding. That is, scientists might arrive at the terms because they give shape to a vague and just-forming epistemology. Subsequently, the terms create images that reinforce or refine that epistemology.

Perhaps metaphor and other language devices gain currency among the postmodern sciences because these literary devices also are means of information exchange. It is reasonable when reading these interviews with Institute scientists to wonder whether new metaphors (ones suggesting a holistic interaction and exchange of information among objects) are emerging to replace the metaphors that portrayed reality as unchanging at its essence, mechanistic, structural, and technological—the metaphors of Newtonian physics. Those metaphors that described the economy in "equilibrium" or molecules as the "architecture" of nature, for example, are being challenged in some SFI research by new metaphors of an economy that is an "evolving complex system" and molecules as "dynamic functions."

In economics, for example, the nineteenth century paradigm (borrowed from physics) portrayed individuals as object-like agents that attempted to maximize their own well being given budgetary constraints, almost in the way a ball would roll on an inclined plain subject to the limitations imposed by friction. This paradigm is giving way to a new information exchange paradigm, developed in the mid-twentieth century by game-theoretic mathematicians. They see the behavior of individual "players" in the economy as constrained by their interactions with other individuals and expectations of what those interactions would mean to each player's well-being. Again, the model of brute force cause-and-effect physics is replaced with a model based on nonlinear information exchange. Game theory is standard methodology at the Santa Fe Institute for all the sciences, including physics. The original mother science of materiality is now a science of information, influenced by the social sciences whose methods it once determined.

Yet in other cases, SFI researchers retain the old physical and technological metaphors of reality; the brain is described as a series of "wires" connected to "nodes," or human society described as following "boundary rules." Arguably, then, neural networks, cellular automata models and other SFI simulation techniques have only appeared to replace a mechanical model of reality with a data-exchange model, but without fundamentally altering the image of reality as distinctly separate from and indifferent to human consciousness. The SFI lexicon is not always representative of contemporary science, nor are most of these terms exclusive to the Institute. Still, it is useful to keep this new information paradigm in mind as we consider what the SFI members have to say about the specific language they use.

REPRESENTING REALITY

The challenge that Santa Fe Institute scientists face in crafting meaningful language to express scientific knowledge is, as we have seen, a challenge of representing reality. It was apparent from the earlier discussion of literal language that the question of whether reality can be represented is a fundamental problem for any epistemologically based philosophic inquiry. Those who believe that science can represent reality through language would be considered realists, "who think that the observable facts provide good indirect evidence for the existence of unobservable entities, and so conclude that scientific theories can be regarded as accurate descriptions of the unobservable world" (Papineau 298). Some philosophers of science hold a restrictive view of realism, known as "monism," which claims that each aspect of reality has just one accurate representation (Longino 44). David Locke writes that scientific representation is premised upon three assumptions: There is a real world to represent, it can be known to scientists, and it can be represented in language (26). (These assumptions do not address the difficult question of whether mathematical expressions are a kind of language.)

Locke and other scholars of science argue, however, that no representations of reality can be possible absent an individual observer, whose observations are colored by his or her own concepts of the world. This recalls the argument of Kant: we order our experiences according to preexisting categories. Evelyn Fox Keller, a scientist and rhetorician, asserts that representations of reality cannot be neutral, free from the observer's pre-existing concepts. "Since 'nature' is only accessible to us through representations, and since representations are necessarily structured by language (and hence, by culture), no representation can ever 'correspond' to reality" ("Secrets" 5). This encapsulates arguments we have glimpsed from Kuhn, Feyerabend, and Latour and Woolgar, which assert that the reality represented by science depends a great deal upon the culture of scientists.

Keller, however, is careful not to let her claims lead to the absurd conclusion that because culture influences our representations of reality, there is therefore no reality, no true structure of nature. "Though language is surely instrumental in guiding the material actions of these subjects [scientists], it would be foolhardy indeed to lose sight of the force of the material, non-linguistic substrata of those actions, that is, of that which we loosely call 'nature'" (33). Keller offers a sensible middle-ground position to the strong rhetoric-weak rhetoric debate among rhetoricians of science, and it is a ground that I believe most of the SFI scientists interviewed would feel comfortable standing on. This ground provides a solid foundation for my beliefs as a scholar reporting on the insights of those scientists.

The problem, of course, is in trying to represent nature and show how it behaves without giving rise to spurious insights that are distorted by observer biases. An instrumentalist view of scientific reality would bypass the problem of tainted knowledge altogether by suggesting that we cannot know reality at its essence, but we can make assumptions that are useful because they are predictive of future events (Papineau 298–299). This assumption is at the heart of positivist science in light of Popper's critique, which holds that even repeated observations are not sufficient to prove a theory's veracity. Instrumentalism challenges the Aristotelian project of trying to determine the essences of matter. As physics evolved in the eighteenth century, its best thinkers moved toward an instrumentalist perspective. Renouncing Aristotle, they would hold, "We can explain nothing in nature completely; we can only derive one phenomenon from another" (Heilbron 72).

Santa Fe Institute co-founder George Cowan is a physicist and veteran of the Manhattan Project at Los Alamos National Laboratories. Cowan (SF) approaches an instrumentalist claim in his interview when he argues that reality consists of more dimensions than any human being is able to perceive. Therefore, he says, "the brain always takes those systems and produces simpler abstractions. . . ." In essence, we make reality useful by operating as if it had a reduced number of dimensions. Cowan suggests that our physiological inability to perceive a world that may have as many as ten dimensions forever restricts our attempts to represent that world in any kind of verbal language that is meaningful. In essence, our perceptions of reality are corrupted by our limitations as three dimensional beings living in what may be a greater than three dimensional world. Scientists try to explain a ten-dimensional world as "strings" that fold over themselves to become manifest as three dimensions, but, as Cowan says, "I don't think the words are very adequate."

Cowan throughout his interview often returns to the concept of "dimensionality," a term that is integral to modern physics and cosmology, but is difficult to define in a colloquial sense. Space has always presented problems in science and philosophy. (For Kant, space and time were conditions that allowed all objects to exist and be known, even though space and time could not be known apart from those objects.) Typically, we think of a dimension as one of three directions—length, width, or depth. We can perceive of time as a fourth dimension. But we would tend to equate dimensionality with the space and time in which we can physically move, collapsing the unknowable into that which he can know. For example, we would think of a line as having one dimension—length—assuming that line is infinitely thin. But if the line is irregular and jagged, it could be seen as having more than one dimension, but less than two. It does not have width, but it mimics width in all its twists and turns. Scientists schooled in chaos theory refer to this as "fractal dimensionality" because the system (in this case our line) has some fractional dimension between one and two. This equates to the physicist's notion of a "phase space," or to the statistician's concept of "degrees of freedom," which influences statistical probability.

The amount of space that a jagged line can move within is greater than the amount of space available to a straight line. Yet, even scientists who expand their understanding of dimensionality to accommodate fractal space would still be hearing the word as a geometrical concept. Hearing additional meanings useful for building complexity theory requires that the scientist not be overwhelmed with impressions of observable space. For example, Cowan uses the term also to suggest the number of variables that control a system's outcome—in other words, the metaphoric space in which a system can operate. If you have a number of people trading on the stock market and none interact with each other or have the same objectives, then the outcome of that market cannot be predicted. Prices would gyrate according to the arbitrary whims of all traders, but without any rule linking their behaviors. The system would have many degrees of freedom, many dimensions, but it would not make sense to an observer.

Complexity theory is interested in systems that have enough degrees of freedom to offer agents choices. So their behavior is not cast in stone; they make those choices by interacting with a limited number of other agents. Notice here how concepts of space, time, freedom to act, and information are present in the term "dimension." A system without information exchange has multiple degrees of freedom (no one knows or cares what the other does); such a system is unpredictable and unknowable. A ten-dimensional universe is unknowable, but it manifests itself to us in manageable three, possibly four dimensions, and in doing so, gains order and predictability.

Nobel laureate economist Kenneth Arrow (SF) in his interview acknowledges that any attempt to represent this strange reality by scientific theory is limited by how much can be known by human observers. Yet, for Arrow—acting as a scientific realist—such representations remain genuine and valuable despite their limitations. "Look, any point of view is restrictive," Arrow says. "That's its value. So I don't think it's anything bad. . . . It's true of any theory that anyone's ever proposed that it simultaneously tells you more about reality to the extent it's true and blocks the development of further thinking." Understanding, for Arrow, requires focus. The "truth is in that restrictive view," he says.

Arrow might well agree that hearing the truth of a theory requires the scientist to tune out some of the harmonic associations of metaphoric terms such as "dimension" in order to hear a coherent, more Platonically beautiful melody. A scientific realist could intuit knowledge of a multi-dimensioned reality by accommodating (in this case, by tuning out) harmonics suggesting that such dimensions must be accessible to human perceptions of physical space. Of course, we could look at this challenge to the scientist another way: perhaps the goal is not simply ignoring loud harmonics, but also hearing soft ones. From this perspective, our realist must learn to ignore the strong harmonics that over-determine a sense of geometric place in order to become sensitive to the weaker harmonics that convey more subtle meanings. In either case, the result would be hearing the term "dimension" in a new way.

Since Santa Fe Institute scientists are very much trying to explain such concepts as they relate to complex adaptive systems, it is clear that they are striving to be scientific realists. Yet, as Arrow points out, they would acknowledge that the explanatory scope of such realism is limited by metaphors and their harmonics. At some point representation by verbal analogue breaks down. Even after hearing the full range of meanings in a word like "dimension," we are unable to envision a reality of more than four space-time dimensions. Invariably, scientists frustrated with the imperfections of language turn to mathematics to extend the scope further toward truth. This can be a productive turn, but, as we will see, it cannot put off forever the need to accommodate theory to unsettled words.

PLATONIC ARGUMENTS FOR THE SUPREMACY OF MATHEMATICS

In accepting the limitations of humans-as-observers, scientists are accepting a Platonic view of the world, argues science writer George Johnson (SF). In his interview, Johnson suggested that most scientists realize that their representations of reality are at best shadows of the truth. He questions whether "any language, including mathematics, is ever going to give you a one-to-one correspondence with reality." But Johnson suggests that scientists believe they can gain degrees of accuracy in their representations and approach the truth by trading words for mathematical symbolism. "They think of something like English or any verbal language as something that evolved through a bunch of frozen accidents," Johnson says, echoing the lexicon of Santa Fe Institute evolutionary theory. That is to say, expressions of discourse catch on in human society and color human perceptions even though those expressions are not in isomorphic correspondence with reality. Language evolved for purposes of human interaction and exchange of ideas, which are not always in concert with true and exact representation of the underlying structure of reality.

Mathematics best represents that underlying structure for scientists, Johnson, says, even if the mathematical constructs create a reality that cannot be comprehended in any normal sense given the limits of human perceptions. Those scientists who are mathematical Platonists are willing to trade true understanding, in a way that would allow them to visualize the picture of reality produced by their theories, for what they see as a more accurate, if shadowy mathematical representation. Hence, Cowan suggests that mathematics can capture aspects of a multi-dimensional universe that are inaccessible to us in ordinary language. "If you are communicating with people who understand your mathematical logic you would use that because you can cram more information into a sparse mathematical concept than you can into a hundred paragraphs of words," he says. In deferring to mathematics to cover the inadequacies of language, Cowan is setting up an interesting dichotomy, whereby a ten-dimensional reality could be represented accurately in a mathematical language, without being understood in the sense that a human being could draw any kind of mental picture of that reality.

Mathematical physicists who have developed theories of a multi-dimensioned universe hope to reveal the link between all the particles and forces of physics, including gravity, and the structure of space. These "string theories" require combinations of the most sophisticated methods of mathematical physics available. They carry odd sounding names such as "noncommunitative algebra," "K-theory," "fiber bundles," and "Calabi-Yau manifolds." Yet, even these theorists recognize that the mathematics have surpassed human perception. The literature refers to problems of "string phenomenology" and asks whether these systems can produce a model that works mathematically, yet also agrees with observations (David Gross 2). Entire conferences in theoretical physics have been devoted to the phenomenological question; in essence questioning whether empirical science will ever catch up with the mathematics. Edward Witten, one of the pioneer string theory physicists, notes in a paper on K-theory that some of the mathematical symbols he uses do not have analogues to reality. For example, he writes: "Again, the physical meaning of the U (N) or E_8 gauge fields is not clear" (4).

These observations about string theories lead to the contentious world of the philosophy of mathematics, where debate has churned since Classical times over whether mathematical insights can be trusted to construct meaning from the physical world. The technical writer who works with scientists invariably will encounter attitudes about mathematical knowledge that have their origins in this debate. It began with Plato and his successors, who argued that mathematical objects were real, although not bound by a physical presence in time and space (Brown 11–15). Recall that for Plato and ancient philosophers of the Pythagorean School, evidence of mathematical relationships could be found in musical scales, the orbits of the planets, and other manifestations of reality.

Hence, most Platonists would argue that numbers are real— even if abstractly so. The concept of a number can be trusted in some way to depict the attributes of reality; it does more than merely offer a way of organizing relationships manifest in that reality. So if a Platonist looks at a carton of six eggs, she sees the set of six and is able to perceive an abstract concept of sixness (Colyvan 5). A nominalist, by contrast, would argue that numbers are less cognitively powerful because they do not exist. Hold up an egg to represent the concept of "one" and a nominalist will reply that he sees nothing of the sort, only an egg.

This debate may seem like a philosophical splitting of hairs, but it lies at the heart of the questions that the Santa Fe Institute scientists raised repeatedly throughout their interviews. In advancing claims that a multi-dimensional universe can only be described by mathematical abstractions, Cowan would not require that the equations allow us to envision ten dimensions. Our imaginations are but clouded mirrors of our limited perceptions. Yet, he and other mathematical Platonists would argue that these equations accurately demonstrate aspects of the universe in ways that cannot be conveyed by words. Hence, mathematics is said to be "indispensable" to science.

This argument is associated most often with the twentieth century analytic philosophers Willard van Orman Quine and Hilary Putnam. The latter summarized the argument in a famous assertion: "(Q)uantification over mathematical entities is indispensable for science . . . therefore we should accept such quantification; but this commits us to accepting the existence of the mathematical entities in question" (Quine, quoted in Brown 53). Quine held that there is no essential "first philosophy" of science that must precede mathematical representation, in essence making the claim that we do not need to first interpret reality verbally in order to represent it (quoted in Colyvan 22). And, more important, once we have represented reality mathematically we must accept the truth of that representation. Hence, it follows that if a mathematical representation of a ten-dimensional universe functions to represent reality in a consistent and coherent way, then the mathematics enjoys "whatever empirical support the scientific theory as a whole enjoys" (10). Brown argues that Quine and Putnam see mathematics describing reality (56), although clearly their description would be highly abstracted from anything that could be perceived.

Nominalists like Harty Field counter that mathematics is not essential to scientific theories, but that it merely provides a fictitious shortcut for understanding what would be possible without numerical equations. For Field, mathematics helps to name aspects of reality (hence the label of "nominalism") the way a novelist might name her characters. He argues from a consequentialist position that the proof of a theory is in the inevitability of its consequences, and that such consequences unfold independently of the mathematics. Mathematics represents reality, but the representation is ancillary to reality's underlying properties. Field champions a program of "science without numbers" in which concepts such as Newtonian gravity can be represented by logical statements that posit consequential relationships between aspects of reality, but not by numerical qualities (quoted in Colyvan 68–75).

As we would expect, Field's program has drawn rebuttal from mathematicians and philosophers who offer examples of scientific phenomena that seem intricately bound to the mathematics underlying them. Philosopher of mathematics Mark Colyvan argues that the mathematics of complex numbers involving the abstract square root of negative one ($\sqrt{-1}$) is necessary to solve differential equations that describe the flow of fluid through pipes and other dynamic activities (82). Colyvan holds, however, that this equation does more than provide solutions to problems in dynamics; it also reveals "deep structural similarities between the systems portrayed by these equations" (83). So mathematics from Colyvan's perspective would allow scientists to better represent reality, which is the goal of scientific realists, and not simply to predict its behavior. The abstract mathematics is necessary to show the unity of behavior across different systems. "Even when the systems are governed by different differential equations, structural similarities may still be revealed in the mathematics" (83).

We have seen Cowan move from an instrumentalist argument, which recognizes the limitations of humans to perceive and make use of reality's many dimensions, to a realist position when he suggests that mathematics can provide representation (assuming, of course, that realists would accept mathematics as a kind of language). His shift is one that has occurred throughout the history of science, particularly in the age of Newtonian physics. Prior to the mid-eighteenth century, physics was the study of hidden or "occult" tendencies of matter that were little understood. Even as he advanced sophisticated new mathematics, Newton worked within this occult paradigm; he used words like "attraction, "impulse," or "propensity" in a way that was seen as sloppy or "promiscuous" (Heilbron 56). Troubled by these words, mathematicians of Newton's day allowed that they should be used only until the underlying causes behind the phenomena were discovered" (59). Clearly they were searching for a mathematical way around the problem of imprecise language and metaphor harmonics.

METAPHOR FOR UNDERSTANDING A NON-ALGORITHMIC WORLD

Yet, philosophers of science who accept that mathematics is indispensable to science must stop short of the claim that it is sufficient to produce understanding. Knowledge that is understandable given the limitations of human perceptions eventually must be expressible in ordinary language, or at least in a mathematical language that corresponds to perceptible reality. Mathematical equations whose symbols have no correspondence with what can be perceived may offer speculative insights and induce wonder, but they cannot alone produce knowledge. Science philosopher Hans Reichenbach writes, mathematical symbols sometimes "have a life of their own, so to speak, and lead to the correct result even before the symbol-user understands their ultimate meaning" (174). His insight is prescient, but it would have been more appropriate to the current world of mathematical physics if it had said, "even if the symbol-user never understands their ultimate meaning."

A claim for the indispensability of metaphor would be based on the premise that we cannot know what we cannot experience. Philosopher of science Sir James Jeans in rebutting the Kantian notion of *a priori* knowledge—that which is present without experience— writes that, "knowledge was found in the human mind, not because it was born there, but as a sort of sediment left by the flow of experience of the man-sized world through our minds" (70). Hence, we could never devise a pictorial representation of reality that is true to the mathematics behind it, Jeans writes, but we can make aspects of truth comprehensible through those pictures. It is only through this process of making reality comprehensible that we produce knowledge, even if that knowledge is at best a shadow of the truth, to use Platonic termi-nology. As Jeans says, in referring to the modern physicist: "He did not see that all the concrete details of his picture—his luminiferous ether, his electric and magnetic forces, and possibly his atoms and electrons as well—were mere articles of clothing that he had himself draped over the mathematical symbols; they did not belong to the world of reality, but to the parables by which he had tried to make reality comprehensible" (16).

For Keller, the interaction of cultural beliefs born out of experi-ence with the non-human material aspects of reality is what leads to knowledge; such interaction is perhaps what Jeans meant by "cloth-ing" the electron. "Only where they [cultural insights] mesh with the opportunities and constraints afforded by material reality can they lead to the generation of effective knowledge," Keller writes (35). In this sense, metaphor and other figurative devices may be articles of clothing that give form to reality by dressing the unintelligible in the fabric of human understanding and experience.

Cowan (SF) suggests that metaphor took hold in epistemology as soon as the natural philosophy of the ancients evolved into the physical sciences of things and phenomena that could be seen or envi-sioned by human beings. Classical scientist-philosophers such as Democritus and Lucretius postulated a material world based on invis-ible atoms of matter, thereby setting up a metaphoric picture of an underlying reality of spherical forms connected by hooks and motes of light—all of which were too small to be seen by the human eye. Metaphor was useful in portraying this invisible reality because those early scholars assumed that reality would look like the metaphoric picture, if only they could see it.

For Cowan, the central model of classical physics was that of the sun as a strong central force governing a field in which other smaller bodies interacted. It is a "wonderful metaphor," he says. His use of the term "metaphor" here suggests that for Cowan, all language that creates a visual image, as opposed to a mathematical interpretation, is metaphoric. Perhaps this orbiting planet image is metaphoric in that in conjures up another even more powerful and ancient image of a mother who controls the behavior of her offspring. As Cowan (SF) says, "You had a strong central force and something revolving around it. . . . It turned out that that metaphor worked for every kind of simple physics you can think of. You and I interact with the earth through its mass, so the earth's gravitational force is all you and I know."

Metaphor, however, falls short in trying to portray the unimaginable reality of modern quantum physics and cosmology because nothing looks like ten-dimensional space-time. For example, Johnson (SF) points out the limitations of the common metaphor now used to depict Einsteinian gravity not as a force, but as a curvature in a universe where time and space are fused into a fourth dimension. We are told to imagine space-time as a rubber sheet and a three-dimensional star as a marble that lies in the thin rubber. The marble "star" causes the rubber to stretch, which would then cause smaller marbles to roll toward it. Yet, as Johnson points out, the picture is tautological; we must assume the existence of a force like gravity in order for the large marble to stretch the rubber sheet.

While Cowan says that we are able to understand the solar system visually as a small round earth orbiting a larger sun, he adds that such a picture could quickly become unmanageable as a representation of local cosmology if we lived in a solar system with two suns—a binary star system. The term "orbit" would lose meaning when three bodies were interacting; which ones would be the orbiter and which the orbited? (Even the idea of orbit between two bodies is problematic, since technically they both circle a common center of gravity instead of one orbiting the other.) This "three-body-problem" also has confounded the deterministic mathematics of classical mechanics. Still, Cowan says that a numerical representation can get closer to the

truth than a metaphoric one. Metaphor is useful as a means of opening the mind to new insights, Arrow (SF) maintains. But he acknowledges the argument often made in science, that "Once it's rhetoric,
well then it sort of lowers its value at getting at accuracy, objectivity,
truth, or something like that."

Biologist Stuart Kauffman (SF) in his interview counters that
narratives of experience are necessary to make reality knowable.
Kauffman is the Institute scientist who seems most comfortable with
using metaphor and narrative to generate ideas. For Kauffman, reality is more than the underlying code that may set it into motion.
Evolution is full of accidental mutations that turn out to be valuable
to a species, hence aiding that species in successful propagation.
Suppose, Kauffman says, that a person had an abnormal heart that
was sensitive to earthquakes, allowing that person to predict an
earthquake minutes before it struck and, thus, take evasive action. If
earthquakes were common, that person would be more successful
and would survive to reproduce in greater numbers than other
people might do, spreading the mutant gene throughout the species.
So even though this mutation had nothing to do with the primary
function of the heart, which is to pump blood, it would still imbue
fitness. Kauffman refers to this is a "preadaptation," a concept that
Stephen Jay Gould made known under the term "exaptation."
Kauffman's point is that these twists and turns of reality cannot be
predicted mathematically, so we are left with stories after the fact to
explain the world. "Somehow or another reality is richer than all our
descriptors," Kauffman says. "So I think that stories are the ways
autonomous agents do, or would, orient themselves to act meaningfully in some situations."

Ironically, as Johnson (SF) shows in his interview and writings,
many scientists whose scientific discourse is full of narrative and
metaphor nonetheless would continue to argue that the important
ideas of science should be adduced mathematically—that rhetorical
devices offer only a rough estimation of reality, not a precise description. Some of these scientists use language with all its rhetorical power
and then seem to want to deny doing so. Johnson's 1999 biography of
SFI resident Nobel Laureate physicist Murray Gell-Mann, *Strange
Beauty: Murray Gell-Mann in Twentieth Century Physics*, tells the story
of Gell-Mann and others on the search for quantum particles inside of
an atom that appear and disappear in a matter of nano-seconds. This
book shows that reality seems more concrete to some scientists and

more abstract to others. For Gell-Mann, "Quarks were 'mathematical,' not 'real,' more like patterns than things, part of abstract symmetries . . . " (*Strange* 226). Gell-Mann's student, George Zweig, envisioned quarks—which he called "aces"—more metaphorically as "something that could be snapped together like Tinkertoys" (226).

Yet, even as Gell-Mann treated his particles as mathematical constructs, he referred to them in poetic language. The word "quarks" is adapted from the James Joyce novel, *Finnegans Wake*. Gell-Mann dubbed the relationship between the particles made up of quarks as "the Eightfold Way," which borrows from Buddhist scripture. "Gell-Mann had learned the power of naming," Johnson writes. "He had not just discovered a series of abstract concepts that helped make sense of the atomic realm, he had bestowed the names that anchored them in people's minds" (*Strange* 233).

Figurative language, including metaphor, certainly aids in understanding. Gell-Mann's contemporary in particle physics, Richard Feynman, gained insights into how subatomic particles called "hadrons," which include protons, might flatten out like pancakes when accelerated to high speeds. Feynman perhaps is best remembered for developing "Feynman diagrams" to help physicists understand elementary particles by envisioning how they interact. These are relatively simple drawings of particles in motion, shown as lines, occasionally connected to each other by squiggly lines that represent the exchange of smaller particles, or information. "Little pictures like Feynman diagrams are always described as being this huge breakthrough not because they said anything differently from what you could say mathematically just with the equations, but with the diagrams people could suddenly really, really understand and grasp these things that were very difficult," Johnson (SF) says. He notes that physicist Julian Schwinger developed similar insights into particle interactions, but for Schwinger everything was "formalistic" and "elegant." Likewise, Johnson, adds, Gell-Mann seemed to think primarily in mathematical formalisms, "and only under great duress translated them into the imprecisions of human language." Johnson's comments suggest a reversal of the oft-assumed path of scientists progressing from metaphor, as a general tool of theory invention, into mathematics for the more precise expression of that theory.

Gell-Mann is well known among the scientific community for his writer's block. In the most telling instance, Gell-Mann was unable to complete the manuscript of his Nobel Prize talk that followed his winning in 1969. Proceedings of the Nobel conference after the awards ceremony list Gell-Mann as a winner in physics, but with a note that he did not present his paper in time for inclusion in the proceedings (Johnson, *Strange*, 269). In his interview, Johnson (SF) suggests that Gell-Mann may suffer from this common impediment because he also is an expert linguist who knows many languages. Knowing the nuances of language may paralyze Gell-Mann in trying to choose the perfect word, Johnson (SF) suggests. For Gell-Mann, "you can just imagine every word is just kind of shimmering with these connotations and historical interpretations." Gell-Mann as a writer is aware of the harmonics of metaphor; hence he would have great difficulty accepting a text that always goes astray with implications.

This leads Johnson to ask if Feynman, Schwinger, and Gell-Mann thought differently or if they just appeared to think differently because of the way that each made his thoughts public. Here, Johnson is approaching the fundamental question of metaphor in epistemology, whether it produces knowledge or is employed after the fact to explain knowledge. Popular accounts of the way Feynman worked suggested that for him, diagramming subatomic particles was a basic tool of cognition. Yet, it is of interest to note that in *Strange Beauty* we find Feynman lamenting that he did not discover anything new, but he merely repackaged ideas that others had discovered.

The Feynman example shows that pictures and figurative language help tidy up fragments of theories and makes those theories coherent both to scientists and to non-experts. Margaret Alexander, SFI librarian, argues that metaphors are the only way to convey the concept of complexity to the general public. "I've worked here a long time, and I can't tell people—my relatives—exactly what is done here," Alexander (SF) says. "Metaphor would be the only way to describe it." Kauffman (SF) makes the same case, stating that metaphors are integral to science not just because human beings think metaphorically, but because the world is not deterministic and, therefore, cannot be tidily summarized in mathematical algorithms. Kauffman's comments suggest that for him, metaphors are valuable because they have more flexibility than do mathematical formulas for dealing with an uncertain, non-determined world. "Since Newton we have been taught to do science by pre-stating a configuration space of possibilities," Kauffman says. Yet, he continues, pre-statements are not possible for living systems. "I think it means that we are persistently doing things that are genuinely innovative that are not calculable ahead of time." If we could calculate the course of life ahead of time, we would deduce our futures, Kauffman says. "But you don't deduce your life; you live it." Metaphors are necessary to understand that life, Kauffman argues. "They are part of our tool kit to make our way in a world that is actually not algorithmic."

Kauffman is making the distinction between science as a set of rules that describe the dynamics of reality and science as a historical account of states of that reality at different times. Physics, with its laws of force and motion, would be an example of the former, while biology, examining the course of evolution, would be an example of the latter. Platonists are more comfortable with the former science because it allows for rules that can posit relationships among aspects of reality.

Perhaps Santa Fe Institute scientists are unsettled about whether mathematics or metaphor or both are imperative because they are caught between the world of rules and that of history. For example, Cowan (SF)—whom we have seen defer to the precision of mathematics in his seemingly Platonic assertions—nonetheless reveals that its very preciseness may be its limitation. Evolution and the historical aspects of reality are the result of chance that cannot be predicted by a rule. The reality we see is not Platonic; rather it is the result of a series of evolved organizations of the physical and natural world that may not be perfect, but that work nonetheless. "There is no Platonic ideal," Cowan says. "There is nothing in nature that represents a Platonic ideal of any sort. It's all simply good enough."

MAKING SLOPPY IDEAS RIGOROUS

Tim Hely (SF), a post-doctoral researcher whose background is in physics and artificial intelligence, suggests that metaphor involves physiological behavior in the brain. "The brain, in terms of metaphor, is taking mental shortcuts to try and allow information that is learnt to be stored and to be used at a later date," Hely says in his interview. Associations between related concepts (for example, "apple" and "orange," "blue" and "sky," or the tenor and vehicle of any metaphor) are linked physically in the brain by connections, called "axons." The connections between like terms are stronger; they would have a higher "weight" or "strength," Hely says, than would connections between unlike concepts. His comments echo the arguments of language scholar Mark Turner, who writes that the brain creates ideas by linking concepts together, much the way an interstate highway moves cars from place to place. Turner argues in *Reading Minds* that the brain is configured in a way that would make knowledge inaccessible beyond the boundaries of corporeal experience, which would rule out knowledge of a ten-dimensional universe.

SFI members recognize that metaphors and other non-metaphoric descriptive terms serve as a convenient and necessary system of classifying their science, even when those terms are vague. Alexander keeps a running list of scientific "buzz words" that she uses when shopping for books to add to the Institute library. The list is always changing, she says. For example, a term like "genetic algorithm" that was once hot is now less intriguing to SFI scientists, in part because it "has passed into, it seems like, the establishment," Alexander says. "You know, we're not supposed to be doing things that are done in other places." Terms related to physics are less popular among the scientists than when the Institute opened in the 1980s, reflecting a shift toward biology as the new foundational science for complexity theory. Hence, a term like "bio-information" that conveys the relationship between information, mathematics, and biology is in vogue. Alexander's use of the colloquial expression "buzz words" implies a certain faddishness in the way research is classified among scientists, but it also suggests that scientists are constantly testing their metaphors and culling those that seem no longer to be useful representations of new understandings of reality.

When Santa Fe Institute scientists discuss particular terms they are using, they seem to do so with a great deal of respect for the harmonic associations inherent in those terms. Yet, as we have seen, the concept of metaphor for these scientists often stands for any kind of figurative and informal language that may be useful for generating ideas, classifying those ideas, and communicating with non-experts, but not as a substitute for formal theory. John Casti, a mathematician affiliated with the Institute, makes the informal-formal distinction clear. A lot of words we use in everyday language have "informal meanings," Casti (SF) says. "Speaking as a mathematician, and that's my orientation, one of the main things that you have to do is often to formalize." This implies that behind metaphoric description is a more formal and detailed explanation of theory. Arrow (SF) uses a dismissive term "slogan" for excessively figurative terminology, although he clarifies that slogans are different from metaphors. Slogans, for Arrow, appear to be the same thing as the catchy phrases referred to in an earlier chapter by Harold Morowitz as "meta-metaphor." Recall that for Morowitz these are terms that are rich with imagery and poor with precise meaning. No one confuses a slogan with formal theory, Arrow says. "We all know a real model when we see one."

Johnson says that many scientists view metaphor as a tool of convenience. Unlike Morowitz and Arrow, perhaps these scientists do not draw a distinction between metaphors and slogans. Yet, it is interesting that various scientists in their interviews refer to precise and popular terminology as if both are used in place of rigorous original thinking. Arrow, as we have seen, uses the term "slogan" to refer to overly catchy, imprecise terminology. Casti refers to highly defined scientific terminology as "jargon." Casti (SF) says terms like "a complex adaptive system" can be defined carefully by scientists and then used as "shorthand" or "jargon" without risk of confusion among those scientists. Hely (SF) says he tries whenever possible to simplify his speech and use as little specific scientific "jargon" as possible. Slogans often are sloppy popularizations, of course, while scientific jargon can serve as a mental shortcut for members of a discipline. Still, both tend to consecrate power in terminology. So SFI scientists recognize, perhaps implicitly, that figurative popular metaphors and precise scientific terminology both have the potential to detour thinking by directing attention away from reality and onto the language itself. Of course, this argument skirts an obvious truth: knowledge of reality is impossible without language.

Still, metaphors that approach the level of mantra or slogan are domineering because they can distract a researcher who becomes so enamored of a term that she or he loses the flexibility to see the world outside of the frame of the metaphor. As SFI theoretical chemist Walter Fontana (SF) says, "You know the potential danger with a researcher is that you fall in love with your ideas and then you become brittle." Other times the metaphors are so catchy that they imply quick fix, real-world applications for a particular theory that cannot be delivered, at least not in the short term. Suzanne Dulle, SFI director of business relations, says that the language of complexity has flooded the popular business-book press, leading to an expectation among some business people that they can spend a few days at the SFI and come away knowing how to save their company money. "And if you put the title 'chaos' or 'complexity' on the front of your book, it seems you can sell many of your books," Dulle (SF) says. Some business managers hearing about self-organization have asked half seriously if they should let the company run on its own. Dulle notes, "I hear companies come in and say, 'All this reading about complexity, does that just mean we're supposed to step aside and let everything self organize?' "

Some metaphors entail greater risks of imprecision than others, Casti says. It seems he is referring to metaphoric terms whose scientific meaning has been qualified mathematically, but ones so rich with harmonics that the precise meaning is impossible to hear. Perhaps the most cacophonous of these terms in use at the Institute and throughout contemporary science is "chaos." In the popular sense, "chaos" implies uncontrolled disorder. In a mathematical sense, however, it suggests sensitivity among systems to initial conditions that, if known, would reveal a hidden order underlying apparent randomness. As Casti (SF) explains: "Chaos is a perfect example of an everyday word that evokes all sorts of images and impressions and subjective feelings. . . . But when a mathematician or a scientist uses the word 'chaos' or the term 'complexity,' for that matter, they usually mean, let's say, the formalized version of that word." Making a common term suitable for formal use requires "sucking out or dramatically restricting in all variety of ways the kinds of interpretations that it has in informal everyday language," Casti says.

Suppressing metaphor harmonics is often easier said than done. Because "chaos" is such a poetic term replete with associations, people with little scientific or mathematical training believe that they understand it even when they encounter it in a scientific setting, Casti says. Even among the "intellectual world" such associations persist, Casti acknowledges. Scientific words often "get stretched beyond their elastic limit . . .," he laments.

Ellen Goldberg, then SFI president and a scholar of immunogenetics, worries that SFI terms are used freely by people who do not know what they mean. "I think we helped that happen," Goldberg (SF) says. "Because I think we fell into this a couple of years ago where we kept using these words and there was no depth. You know, we were doing very well out there in the lay press, but I think many of our peers out there—the scientists—kept saying, 'Where's the beef?' " Goldberg says that members of the Institute have to do a better job educating the public about the meaning of terms like "edge of chaos," "complexity," "emergence," and the like.

Of course, it is obvious even to a casual observer that the Santa Fe Institute owes its success largely to the metaphoric language that its scientists have helped to spread through the popular culture. The fact that these terms are catchy, accessible, and replete with harmonic overtones allows them to diffuse in a way that could not happen with less evocative terminology. A term with many meanings and implications can sound good in many situations, thereby ensuring its ability to take hold in many discourse communities. Santa Fe Institute scientists have been successful at publishing popular books and at establishing consulting businesses because their language is memorable. To wit: Murray Gell-Mann's car license plate is one word, "Quark," suggesting how valuable this literary name for a mathematical concept has been to his career.

Still, popular language comes at a price. Colloquial terms used metaphorically in science may lull the layperson into a false sense of understanding. Terms like "complexity" or "chaos," for example, transport associations from everyday usage across to a word that scientists may be trying to use with greater precision. Alexander speaks about people coming to the Institute, almost lured by the catchy phrases into thinking that answers to problems in their businesses or other life endeavors will be found easily. "I feel sorry for the people who don't find answers here," Alexander (SF) says. After they read some of the books or attend a lecture and encounter the difficult mathematics behind such seemingly accessible theories, they realize they need more background, she says. So the mathematics acts as a kind of reality check on the metaphors—a wall. "But I think maybe that's an important wall," Alexander adds.

The comments from Arrow, Casti, and Alexander suggest that terms of science that also have broad popular uses could be reined in by use of formal models and mathematics. In this sense, the mathematics makes sloppy language rigorous. For Kauffman, however, it is often the metaphors that illuminate meaning and hone precision. A metaphoric term evokes images and associations that scientists either can accept or reject. This process of building a representation of reality from figurative language spawns other figurative language, serving as a kind of linguistic inductive method for Kauffman. A concept like "the edge of chaos," for example, is partly scientific and partly rhetorical, Kauffman (SF) says. Kauffman and others at the Institute use the edge of chaos metaphor to create the images of biological, social, and physical spaces that lie between complete, unchanging, predictable order and randomness. This is a region where interesting phenomena happen, such as evolution of life forms, a market economy, or climate. Kauffman has a mathematical way of representing such a region, using the symbolic logic of Boolean algebra. But metaphor is necessary to enrich the logic with real examples. "The work that the metaphor will do in this . . . context will be in discovering how it (the logic) is true." Physicist Chris Moore (SF) says in his interview that while some people may think in pure abstractions, "most need language." He adds: "We couldn't do what we do without metaphor."

Metaphors, Kauffman says, aid scientists in developing new ideas. An evocative metaphor causes scientists to "carve up the world into new concepts that then take on real empirical meaning." Echoing the Kuhnian notion of paradigm shifts, Kauffman says that new terms in science may be vague, but they spawn research that helps define their meaning:

> You know what often happens in science . . . is that you get an idea, a rough idea, and it takes decades to figure out what the idea really means. I mean examples are entropy and energy from the . . . [nineteenth] century. And right now it involves ideas like 'autonomous agent.' Okay, just because I thought up the idea doesn't mean I understand it. I've written a whole book on it.

Implied in Kauffman's argument is the claim that language propels the researcher to make fresh associations, which can then be tested. Comments from Goldberg (SF) also suggest that metaphors function inductively in science by leading researchers across a field of various definitions. This semantic exercise eventually refines the meaning that scientists are groping for in a specific context. Goldberg says she is not troubled by multiple definitions of a term like "complexity" or "rules" because the interaction of various definitions spawns new understandings. Multiple definitions among scientists from different disciplines "leads to confusion," Goldberg (SF) acknowledges. "But that increases the communication, you would hope." For example, Goldberg acknowledges that she would not likely have joined the Institute and its intellectual dialogue if she had not been lured by colorful and evocative terms like "complexity." As long as researchers treat the terms with respect and use them thoughtfully, Goldberg says she welcomes the different definitions of SFI terms.

While Casti (SF) argues early in his interview that ambiguous terms at the Institute must be formalized to be useful in science, he later maintains that certain vagueness is useful—especially when building models for new theories. Casti suggests that terms like "complexity" or "sand pile catastrophe," with all their harmonic associations, can inspire insights in the way that an impressionistic painting leads to insights for art lovers. These insights would not be available from a more realistic photograph. Formalizing a metaphor by imposing exacting definitions means that it is not a metaphor anymore, Casti says, but "the real thing." Hely (SF) agrees that vague terminology often is appropriate because it recognizes that the science being described is still poorly understood. For example, in brain biology, the terms "weight" or "strength" are used to describe the power of associations in the brain, as evidenced by links between neurons. But scientists such as Hely recognize several interpretations for the terms; they can refer to the number of connections one brain cell makes with another, the probability that the second cell will pass on information received from the first, the volume at which a cell sends a signal, or even the physical distance between cells.

We continue to encounter the debate over whether language based in mathematics or in words offers more precision, and whether such precision is useful for knowledge creation or whether it is stifling. Kauffman and Goldberg—maybe even Casti—would seem to hold that mathematics might be more precise, but that such precision does not offer the cognitive power inherent in a word-based language rich with associations. So metaphors and other words invite new ideas that are inherent in the harmonic associations of those terms. Rhetorician of science Donna Haraway, however, makes the opposite assertion: word-based language reins in meaning by making it concrete, while mathematics remains abstract. She writes: "The concrete nature of models, metaphors, and artefacts, is essential to science because it limits the implications of any particular abstract system. A set of mathematical relations and operational terms of a well-developed science can be dangerously overextended" (*Crystals* 189).

Haraway's rebuttal to the belief in the supremacy of mathematics asserts that because mathematics is abstract, it can wander far from reality without such wanderings being obvious. Words, however, may have less abstract freedom because they are interpreted in ways that are restricted by commonplace associations. The implication in Haraway's argument is that knowledge creation is as much about restricting meaning and associations as it is about permitting fresh unexpected

associations to emerge. Recall from earlier in this chapter that Arrow
(SF) also argued that all theories by definition restrict meaning.

The term "fitness landscape, " often used at the Institute, offers
insights into the question of how a mathematical language and a word
language differ in the way they create and restrict knowledge. A fitness landscape describes the presence of various organisms (or individual genes in one organism) that coexist at different levels of success in
the same environment. The concept is extended to cover a variety of
evolutionary outcomes in biology, where commonplace associations of
"fitness" and "landscape" imply lifelike entities in some kind of environment. It would be less likely, however, that these words could be
extended into cosmology. The harmonics generated by "fitness" and
"landscapes" sound discordant next to language describing the raw
chemistry of deep space. Yet, consider the following mathematical
expression for fitness landscapes (Kauffman, *Origins of Order* 42):

$$W = \frac{1}{N} \sum_{i=1}^{N} w_i$$

where W is the fitness of the entire "landscape," or the entire genotype, as determined by the individual fitness, w_i, of each member of
the gene community. This formula, where Σ indicates the sum of the
parts, is nothing more than the equation for taking an average. It is
common in mathematics for a variety of arguments and not intrinsic
to evolutionary biology. Clearly, even if each symbol is defined, it is
easy to see how such an abstract formula could be transferred to any
domain of science where averages matter—presumably including cosmology—that have nothing to do with living environments.
Following Haraway's thinking, then, the mathematics permits abstractions that would never be possible with language because words force
associations that symbols like Σ do not. Of course, a mathematical
scientist could argue that if the words "fitness landscape" have less
freedom than do the equivalent formulas, then those words have less
power to capture the wisps of Plato's shadowy world.

Haraway's insight reverses the normal argument in science that
we have glimpsed throughout, which suggests that words are sloppy
and must be rendered more precise via mathematics. She argues
instead that abstract mathematics can be overextended unless mitigated by words. Haraway's argument makes sense when we look at the
history of physics to find eighteenth century Jesuit scholars preferring
mathematical representations of reality to those based in language.

The mathematics allowed Jesuits to avoid the implications of new theories of the solar system with the sun at the center (Heilbron 104). The term "heliocentric" means "sun centered," which simply is a statement about the physical structure of the universe. But unsettling harmonics implied that the jewels of God's creation, humans on earth, were no longer so important. Jesuits preferred to take their heliocentrism mathematically, where they would not be troubled by such dissonance.

Of course, Haraway's argument assumes that equations do not have harmonic associations and, therefore, can be used in any context without sounding discordant. Largely this is true, although some mathematical expressions have strong associations. Take the symbol pi (π), which always implies a circular shape or cyclical pattern of dynamics of a system or an element within this system. A mathematician who found the need for π in her equation would be forced to consider the cyclical properties that it implied. Perhaps words and mathematical formulas can spawn harmonic associations, although such harmonics would be less common in mathematics.

NO RATIONAL METHOD OF HAVING GOOD IDEAS

The larger issue is whether harmonic associations possible in any language are helpful or distracting. Arguably, the uses of multiple languages in science, including mathematical languages and word-based languages, serve as a checks and balances system. The constant struggle between the need for the intellectual freedom inherent in abstractions and the need to validate those abstractions through concrete associations is what allows theories to emerge and then be tested. Otherwise, we are left with a world in which scientific inquiry can at its best produce shadowy images of reality.

Yet, are these Platonic shadows the most we can hope to see? Scientists have always accepted surreal presences such as ether, phlogiston, and even curved space as necessary to make their theories work out, even if they later abandoned those presences when more accurate understandings ensued. Since those shadows are useful "instruments" for predicting future events, perhaps we must accept that reality may be unknowable or vastly different from the pictures we create to render it understandable. The underlying reality of computer software code, for example, appears as nothing like the icons that users see on their screens, although correspondences link the code and the icons.

Scientists acting as instrumental users or manipulators of knowledge may not be as troubled by the inability to clearly perceive the underlying nature of reality as philosophers acting as realists would be.

Arrow (SF), for example, seems comfortable with a Platonic reality whose forms are chimeras at best. For Arrow, whether something "that a practitioner might naively think is a description of reality turns out to be a metaphor or a rhetorical device . . . has nothing to do with the question of whether it's good science." Reality need not be meaningful, or understood to the average person, just as software coding need not be meaningful to the average computer user. Of course, computer scientists can understand the underlying computer code; not even the most gifted physicist can truly understand or envision a mathematical reality of ten dimensions. So the mathematics may be true, from a God's-eye perspective, and absent the naiveté of words, but not meaningful from a human perspective. This insight is consistent with a comment often uttered by scientists that if there is a god, that god must be a mathematician.

As we have seen, such mathematical Platonism is forever in the background of any scientific endeavor, including those that bring writers together with scientists for the purpose of developing and explaining knowledge. This awe of mathematics is widely accepted in the sciences despite challenges from Field and other nominalists who see mathematics merely giving name to reality that is perceived more intuitively. A technical writer working with scientists will have more success if he accepts the indispensability argument, if only to agree that science in some cases requires the logical syntax of formulas. These posit consequences over time among attributes of reality, which have measurable or definable relationships to each other.

Where that writer can draw the line, however, is at the implication that this mathematical Platonism implies supremacy of mathematics over metaphor and other devices of rhetoric. Because mathematics is essential, it does not follow that mathematics is superior to other means of adducing knowledge. Furthermore, the distinction between whether numerical representations of magnitudes and quantities are necessary or whether mathematical statements function more as logical syntax is unclear, as is the answer to whether mathematics describes or represents reality.

Arguments for the supremacy of mathematical knowledge would appear grounded in a scientific elitism akin to the Classical elitism of Plato's school, whereby only the enlightened are privy to the truths—the pure Forms—of reality. Yet, philosophers, rhetoricians, and other humanists are equally unconvincing in defending their ground by attempting to show that knowledge represented mathematically is discursive knowledge in disguise, that science is all rhetoric. Little is to be gained by prolonging a debate over which type of knowledge making is superior.

We can close this discussion about mathematical and verbal knowledge by examining two historical examples from science that show the impossibility of discerning where one type of knowledge leaves off and the other begins. First, consider accounts of the nineteenth century physicist James Clerk Maxwell, who pioneered electromagnetic field theory and thermodynamic understandings of the behavior of gases. It is the field theory that is particularly relevant to this analysis of mathematical and verbal knowledge; in postulating the nature of electromagnetic forces, Maxwell drew from both epistemological systems. Brown's account shows how Maxwell invented the concept of electromagnetic fields purely to represent mathematical relationships between charged particles (50). Maxwell also depicted these particles and fields metaphorically by drawing hexagonally shaped particles, called "vortices," which were linked by small circular "wheels" rolling between them. The diagram made it possible for the mathematical relationship to make sense perceptually, even though Maxwell did not think such a picture was true to nature (Purrington 68). Here we see a scientist reverting to abstract imagery as a way to give shape to abstract mathematics. Electromagnetic fields were shown to be more than mere abstractions, however, when research proved that one particle's movement caused a reaction at a later time in another particle (Brown 50). Cause and effect action over distance and time meant that something was doing the action; in this case it is an electromagnetic field. Both the mathematics and the metaphoric model proved indispensable to Maxwell's theory building. The two complimented each other.

The discovery of subatomic positrons in the 1920s by particle physicist Paul Dirac offers further evidence of the two systems of scientific thought working in tandem. Ironically, it also offers evidence both to rhetoricians and mathematicians of the indispensability of their individual epistemological methods. Philosopher of science and scholar of scientific rhetoric Mary Hesse argues that Dirac envisioned the quantum world of particles in terms of "holes (vector spaces) and jumps in state" This image led Dirac to postulate the presence of some positively charged particle (later called "positrons") even when he could not recognize the evidence for such particles in his mathematical equations (quoted in Hoffman 411). Colyvan, the philosopher of mathematics whom we have been following in this chapter, offers a different account of Dirac's thought process. He argues that his formulas predicted the positron even though the equations did not appear to provide a physical representation of reality. "Thus, Dirac, by his faith in the mathematical part of relativistic quantum mechanics and his reluctance to disregard what looked like non-physical solutions, predicted the positron" (Colyvan 84).

A cursory look at historical accounts of Dirac's research in particle physics reveals that he developed an equation to unite theories of Einstein's relativity and quantum physics, but that this equation required a variable that could only represent a positively charged electron. Rather than accept the paradox that this mathematics entailed, Dirac looked away from the equation to a model in which this mystery particle would be seen as the absence of an electron at an atomic energy level where one normally would be. This was the metaphoric "holes," the existence of which forced Dirac to acccept "a new kind of particle, unknown to experimental physics" (Mehra 48). The equation led Dirac to see a new metaphoric representation of the atom, which in turn made the mathematical paradox sensible.

Johnson (SF) argues that metaphor functions as a type of catalyst that aids in extracting meaning from mathematical statements. Dirac's metaphoric "holes" would carry the harmonic association of a void or empty space—an association that Dirac would have to accommodate, and ultimately reject, in order to envision a positron that was anything but a void. For Johnson, the interaction of these metaphor harmonics with mathematical formulae is the source of scientific knowledge. "I guess I suspect that metaphors are crucial tools to thinking," he says in his interview. "I suspect that even the most abstract physicists really do think in metaphors, if not primarily, at least kind of going hand-in-hand with the mathematics. As soon as they write down an equation, they ask what it means."

Obviously Dirac engaged different intuitions, which arrived in his mind both mathematically and metaphorically. We cannot reconstruct Dirac's thought process with a view toward arguing the supremacy of one system of thinking over the other. Both were necessary and intertwined. Philosopher of science Hans Reichenbach would appear to recognize that novel insights cannot be conjured up merely by following a formal system of thinking, regardless of whether that system is mathematically rigorous or metaphoric. As Brown paraphrases Reichenbach's argument, "there can't be a rational method of having good ideas" (22).

SFI METAPHORS AS THEORY CONSTITUTIVE OR LITERARY? A TALE OF TWO TERMS

If we assume that knowledge is impossible without understanding, then rhetoric and its figures must have some knowledge production role. Metaphors, as we have seen, function as a kind of spotlight that illuminates reality and focuses scientific attention on aspects of that reality in a way that makes it coherent. When scientists attend to any aspect of reality, they usually find that repeated experiments are necessary to assay verifiable facts. They can represent those facts through mathematics or linguistic-based description. By then, of course, the metaphors already have done their work.

It is worthwhile to look at specific metaphors at the SFI, with the assumption that not all metaphors function in the same way. Some may be up to the challenge of constructing scientific knowledge, while others may be more flash than substance. Several of the researchers I interviewed talked at length about specific metaphors they use; they offered a range of opinions about the cognitive function of those metaphors. In this section I focus on two SFI researchers who have differing views on the value of metaphor in their work. One, a theoretical chemist, has developed a new metaphor that portrays molecules as a sequence of interacting mathematical functions. That scientist clearly sees the metaphor as theory constituting. Another researcher is working on ways of discerning patterns of causality in physics. Yet, he argues that his use of the term "pattern" is not metaphorical. Even if it were, he maintains, the use of metaphor in science is at best temporary "scaffolding," which serves until real theory can be built. Let us turn our attention now to these two researchers and their work and consider whether they are using their metaphors differently.

Recall that the word metaphor has at its roots the Greek word *phora*, which can mean "transference" or "locomotion." Associations —what I have been calling harmonics—move from one term to another. Fontana (SF) in his first interview offers similar insights about a kind of cognitive locomotion in SFI metaphors. "The power of metaphor, before it becomes theory, is that it makes people recognize that certain old questions can be cast in new ways," Fontana says. This "triggers new thoughts and speculations." Fontana's use of the word "trigger" implies that metaphoric thinking stimulates new images across the brain.

Yet, this locomotion for Fontana and fellow researcher Leo Buss does not just mean the substitution of words for each other. Fontana and Buss have developed a way of envisioning chemistry whereby dynamic mathematical functions are used to represent molecules of interacting elements. Molecules, we recall from high-school chemistry, are the smallest units of a substance that retain its unique properties. Traditional science often portrays molecules as tiny isolated bits of recognizable substances that have fixed properties, almost as blocks used to build the physical and natural world; chemistry is when these bits of matter are joined with other bits in a reaction that creates compounds with new attributes. One college chemistry textbook first published in the 1940s, for example, conveys the modernist view of molecules as "the structure units, the islands in space, of the states in which the substance may exist" (Timm 21). A 1964 book for young people by Linus Pauling, with drawings by Roger Hayward, shows the standard image of a hydrogen molecule as a dumbbell shape, with two spheres rigidly connected by a center bar. The title of this book, *The Architecture of Molecules*, reinforces the metaphor of molecules as stable building blocks. So, for example, when the units of what we recognize as salt, made up of atoms of sodium and chlorine, are joined with units of magnesium sulfate, made up atoms of magnesium, sulfur, and oxygen, we end up with the white powder, sodium sulfate, useful in making glass. The building-block image implies a builder—perhaps God, the scientist, or nature—who assembles those blocks into the superstructures of reality.

Fontana and Buss, however, envision their molecules as being variable objects that change fundamentally as they link with other molecules. Each molecule is treated as a dynamic object that can be represented by a special type of calculus, "Church's lambda calculus," which blurs the traditional boundaries in mathematics between variables and the functions that direct interactions among variables. In lambda calculus, a function in one stage of the calculation becomes a variable in another stage. So the molecules in chemistry are themselves "functions," as Fontana (SF) says, not just isolated units of a substance that are moved around in a function. The molecules in this metaphor are granted the agency that is necessary to interact; molecules now act, and they are no longer dumb participants in a reaction initiated by a higher order of reality. "You can see that this is an inductive step," Fontana says, suggesting that he and Buss arrived at their metaphoric approach to chemistry after a careful step-by-step process of reasoning supplemented by empirical observation. The metaphor obviously did not appear in a eureka moment, but was the product of painstaking intellectual exploration. "I take the analogy seriously, the metaphor seriously," Fontana says, adding later in the interview that this metaphor has been recognized by many scientists as "an oblique, original way of thinking."

In this metaphor, molecules gain agency by being able to function. In essence, the molecules are tiny computer programs, carriers of information in the form of "rules" that tell them how to function when they interact with other molecules. Johnson, in his book *Fire in the Mind*, describes how Fontana simulated this kind of molecular activity on his computer by representing the molecules as geometric shapes that interact to create new shapes. Interaction begets new forms. As Johnson writes, "Three circles might invariably be followed by a square. A grammar had emerged. Physics gave rise to chemistry" (*Fire* 250). The old metaphor of the parts of matter as dead building blocks gives way to new metaphors where matter becomes information that is able to communicate with other vessels of information and generate its own higher order of reality. A molecular reaction, says Fontana (SF), "is just plain syntactical manipulation."

Fontana's metaphor of an object as a function clearly is theory constituting, not a mere literary ornamentation. He says that researchers could have developed any number of "little private chemistries to mimic particular chemical systems, but none of those systems would give you any insights." A non-metaphoric "private chemistry" would bypass the metaphoric transference of functionality across to an object—sacrificing the essential cognition in exchange for perhaps a more precise formula. The metaphor, however, is not entirely satisfactory in its present form, Fontana continues, because the lambda calculus does not explain the method of interaction and organization that he and Buss see as the basis of chemistry. So they must modify the off-the-shelf calculus "to do justice to certain aspects of chemistry," Fontana says. Others might agree that the lambda calculus metaphor of chemistry-as-a-function lacks the precision needed to ground a robust scientific theory. As Fontana says, "Some people tell us, 'Look, that's very nice, but you know . . . what I don't like about it—it is too general.'"

Per Bak, a physicist associated with the Institute, has argued that complexity theories by nature must be abstract. If so, a lambda-calculus model of chemistry might work because of its generality, not in spite of it. SFI theories follow a systems approach to science; this looks at overall patterns of interactions among agents, not always at specific agents. In a 1996 book, Bak writes that a theory of life should be able to describe the mechanism for life on Mars if it were to occur. Such theories must be statistical and not obsessed with specific observed events. This kind of science has moved away from Enlightenment empiricism, suggesting that attention to individual details causes one to lose sight of the big picture. "If, following traditional scientific methods," Bak writes, "we concentrate on the accurate description of the details, we lose perspective" (10). The metaphor of chemistry as a function seems to satisfy this requirement for abstraction because it removes focus from a specific chemical reaction and looks at the process of chemical reactions. As Fontana (SF) says, "It is general because the phenomenon is general. It's not limited to chemistry alone."

Metaphors in science, even when carefully worked out, as Fontana's obviously was, usually do not form a perfect fit with the evidence. As we recall, Boyd maintained that metaphors have a heuristic function; they offer "epistemic access" to partially formed ideas "until more precise accommodation is achieved in the light of later discoveries . . . " (399). This process of accommodation occurs when the schemata of the brain, perhaps structured to fit a specific metaphor, are modified in response to new information. Fontana recognizes that his metaphor will require some adjustment, yet the inexact fit in no way detracts from his appreciation of the metaphor's knowledge-creating value.

Cosma Rohilla Shalizi (SF), a statistical physicist, in his interview makes similar observations about the heuristic function of metaphor, but argues that the value of metaphor in creating knowledge is at best temporary. "It obviously is very useful in communicating things to people . . . to help build up that knowledge in people, or even in yourself," he says. "So it's constructive, but not in the usual sense people talk about constructivism when they're talking about knowledge acquisition and education." Shalizi describes metaphor as "scaffolding, or some temporary structure which we use to erect something more permanent and perhaps more integrated." Shalizi's use of the word "scaffolding" has precedence in science. The nineteenth century physicist James Clerk Maxwell made a similar reference in describing the ways in which Andre-Marie Ampere developed his theory of electromagnetism. Maxwell wrote that Ampere " . . . removed all traces of the scaffolding by which he raised it" (quoted in Purrington 47). Shalizi says he is skeptical of claims that science is mostly rhetoric, or that metaphors construct knowledge and determine how scientists think.

Metaphor for Shalizi seems to be a useful tool, but one that the scientists control—not the other way around. He acknowledges that theories by people like linguist George Lakoff, suggesting that metaphor dominates human discourse and thought, are "plausible." But he still says that metaphor in science is mostly a means of delivering knowledge, not of finding it. "I'm not really comfortable with basing things on plausibility," Shalizi says. "If you do that, you might as well believe Freud." The latter comment echoes Popper's critique of the theories of Freud and Marx as being pure conjecture because they would be impossible to prove wrong.

Yet, Shalizi uses metaphor freely in his writing with physicist James Crutchfield. The two are attempting to discern statistically significant patterns in physics that would indicate that one state of reality often leads to another. This is the question of causality, one of the most challenging problems of metaphysical science and philosophy. Shalizi and Crutchfield in their SFI working paper, "Computational Mechanics: Pattern and Prediction and Simplicity," are careful to examine the semantic aspects of the terms they use. One section, for example, differentiates patterns from classifications and configurations by referring to a passage by novelist Jorge Luis Borges that also begins Foucault's work on classification, *The Order of Things*. Classifications for the two SFI authors are human divisions that help us to understand the world, while patterns are natural regularities that may not be informative to observers (Shalizi and Crutchfield 3). A pattern could be a kind of "rule," Shalizi (SF) says, if you defined rule to be an "effective procedure"—where if you know what X is, you can predict a range for Y.

Shalizi acknowledges that one of the main challenges he and Crutchfield face is in finding a "precise sense" of the meaning of the word "pattern" in computational physics. It is different from "pattern" in biology," Shalizi says, echoing the debate over the term "rules" among political scientists and biologists. A key difference, as he explains it, is that in biology, pattern seems mostly to do with spatial elements—repetition of structure across three dimensions. "Then you could look at something like tissue and you could say, 'we've got these chemicals here, or tissue, or like this sort of cell here,' " he says. "And you could use the analogous techniques to produce a description of that pattern; it wouldn't be causal in the same sense." In computational physics, Shalizi says, a pattern relates more to a description that reveals a "compressed regularity" of causal events in time. Ours "works for one-dimensional systems in statistical mechanisms," Shalizi says. "We don't yet have a good way of extending it to higher dimension systems."

For his research, Shalizi searched library texts for uses of the word, finding for example that "pattern" in art means "geometric designs." Shalizi, however, argues that the way he and Crutchfield are envisioning patterns in statistical physics is not metaphoric. "By now it's sufficiently abstract that it's not metaphorical anymore," he says. "I'd say the same is true of, say, 'structure.'" A "pattern" for these physicists is when a set of equivalent "histories," or conditions of physical reality that all have the same chance of producing a particular distribution of outcomes, do produce with statistical regularity such a distribution. An equally elusive term that appears metaphoric, but likely would not be so in Shalizi's assessment, is "epsilon machine." Shalizi and Crutchfield refer to an epsilon machine to represent the mathematical function that determines when states of the world, the "histories," become causal states.

Shalizi (SF) says that the term "pattern" is useful even though it is vague because it has enough overlap with colloquial associations, but it also can be defined mathematically. So it is not "jarring," he says, acknowledging that even among scientists, familiar, tuneful language often is necessary. Keller makes the claim that scientists like to borrow colloquial terms and imbue them with a technical meaning because the scientists can benefit from the associations that such terms invoke, without having to assume responsibility for those associations. "The use of a term with established colloquial meaning in a technical context permits the simultaneous transfer and denial of its colloquial connotations," Keller writes (121). Using a word like "pattern," which sounds like a metaphor even if its authors do not intend it to be so, for example, may comfort readers of a statistical physics paper—and provide the encouragement to continue reading—by suggesting that what is being discussed is not so rarefied as to be beyond understanding.

This strategy of using a term and benefiting from its popularity, while at the same time attempting to deny that popularity, has its roots in Classical rhetorical strategies. Using a term like "pattern" with the caveat that it has a more rigorous meaning than one would infer from common associations could be seen as a kind of rhetorical *aposiopesis* (*Rhetorica ad Herennium* 290–291). Here a rhetor introduces an idea to an audience but stops short of elaborating upon it, achieving the psychological affect of arousing associations in the audience without having to claim full responsibility for those associations. The strategy most commonly was used when a rhetor intimated some wrongdoing by a rival, but then caught himself and refused to say more—thereby arousing audience suspicion. A scientist who uses a metaphor knowing its harmonic associations can attempt to stipulate away the associations, hence appearing at once accessible and consistent with tradition, but also rigorous and above the popular lexical fray.

Shalizi (SF) says he and Crutchfield might be comfortable coining a neologism for what they are now calling a "pattern," although coining new terminology would be a step of supreme confidence. "To be honest, I'm not sure that we have enough precision down in here to warrant that," he says. In the interview, I suggested the technique used by philosophers for discriminating between different meanings of a complex word by using subscripts (T_1, T_2, for different kinds of truth, for example), but Shalizi says physicists would not be familiar with such lexical markers.

Shalizi's comments suggest that theoretical scientists think of metaphors as images that correspond to the observable world or the mind's eye, just as the term "pattern" corresponds to the geometric shapes that make up patterns in clothing. The need for such pictorial verisimilitude, then, would render metaphors insufficient for fine-tuning abstract theoretical science that defies pictorial representation. This essentially is the argument that abstract reality can be expressed accurately only though mathematics. Interestingly, examples offered by scientists of where metaphor falls short often have time as a variable—whether it be time as a fourth dimension that is manifest as curved space or time as the medium in which statistically relevant patterns of causality play out. This is consistent with the work of Reichenbach who, you may recall, argued that humans define time, but do not know what it is.

It is also notable that while Shalizi accepts the figurative term "pattern" to describe the physical state he and Crutchfield are researching, he seems to want to suppress the harmonic implications that suggest similarities in meaning shared with the word "pattern" in biology. As Haraway shows, "pattern," along with related terms like "form" and "field," comprise the new lexicon of a late nineteenth and twentieth century genre of evolutionary biology. This new genre eschewed the Newtonian-influenced paradigm of a living body as a machine made up of small operating parts, where each part was seen to have evolved solely to function in the final product—the complete organism. Instead, these "organicists" averted attention from the final organism and examined life as the emerging outcome of relationships and interactions among components at various stages of the organism's development. The structure of the parts of an amphibian zygote's nervous system, for example, can be seen as a series of evolving patterns that have direct influence on the next stage of the animal's development, rather than as a schemata of parts that function only after the animal is fully formed and animated (Haraway 67–68). This organicist view, where life is likened to a quivering web of interacting components is embedded in the concept of emerging organization. A shift away from mechanistic metaphors to more organic ones defines Santa Fe Institute complexity sciences, yet the shift has appeared elsewhere in twentieth century science. As rhetorician of science Mary Ellen Pitts notes, for these new scientists, "ours is a world that consists not mechanistically of separate parts to be identified and analyzed, but interactively in a web of ongoing processes" (249).

It follows from this organicist perspective, therefore, that patterns in biology contain an element of causality, meaning that the structure of an organism determines its function. But, as Shalizi suggests, a pattern in biology is three-dimensional and shows up in the tissue and fabric of living material. In physics a pattern has only the dimension of time; it is the relationship among histories. So a pattern in physics is similar to one in biology, but it also is different. Accepting the metaphoric transference of associations across the boundaries of physics and biology invites connotations for both fields that would seem to be useful in the production of knowledge, even if those connotations must be refined later to allow for differences.

Both the scientists highlighted in this section appear to be using metaphor, although arguably for different reasons. Fontana (SF) clearly has carefully developed a metaphoric association whereby the tenor, a molecule, is transformed into some new phenomenon by means of the vehicle, a mathematical function. The molecule is a "frame," to use Black's terminology. It surrounds an idea that changes "focus" when portrayed metaphorically as a function. Fontana's metaphor obviously is not completely colloquial; many non-scientists might struggle to recall what a "function" means in mathematics. But it does transfer the scientist's common knowledge about mathematics to an unusual target—a molecule. The transfer is intentional and essential for understanding the theory that Fontana and his colleague have developed.

Shalizi (SF) also is using a word that appears to be a metaphor, although his intent is not primarily to transfer associations from the word "pattern" across to his statistical process. He clearly recognizes the metaphoric aspects of the word "pattern," and he addresses those aspects in describing how his term in physics differs from the same term in biology. So the metaphor is doing its work, by carrying insights across terms, even if Shalizi does not find such locomotion useful for the fine-tuning stages of his research. In that sense he may be astute in saying that the word, for him, has become an abstraction, or scaffolding. A neologism might work just as well if it could be accepted, although it would sever any cross-disciplinary interactions that may still offer surprising insights in his research.

Shalizi argues that he is using his metaphor less to gain the cognitive insights afforded by metaphoric locomotion, and more to appropriate a familiar term that will make readers feel at ease as they wander into abstract theory. Shalizi acknowledges that figurative language is necessary to help entice young people into the various fields of science. This recalls Goldberg's (SF) earlier comment that she was drawn to the Institute, in part, because of its evocative language. Metaphoric language is essential to make such theories appear compatible with ordinary human means of understanding, through imaging and representation.

The Fontana and Buss metaphor of a molecule as a mathematical function and the Shalizi and Crutchfield metaphor of causality as a pattern are closely related—perhaps not in the goals each pair of scientists had in using the terms, but in the implications each metaphor brings to the new paradigm of information-as-materiality. Molecules in the model Fontana describes are seen not as objects, but as chains of formula that change as they engage other chains. Fontana (SF) reveals just how much he sees molecular structure as a unit of information when he describes it as "syntactical structure that is given by the combination of atoms." Fontana and Buss needed a theory to connect structure with action, Fontana says. The key to that theory was to "rationalize or formalize molecules as almost linguistic entities," in other words, as information rather than as objects. We may need to switch our stance to see molecules as "agents of transformation," Fontana says.

Shalizi and Crutchfield have postulated their "epsilon machine" using mechanistic language for causality. Despite these powerful harmonics, the "machine" is not mechanistic at all. Instead, it is a statistically meaningful process that connects events. Epsilon machines, Shalizi and Crutchfield explain in an article submitted to a journal of mathematical physics, "reveal, in a very direct way, how information is stored in the process, and how that stored information is transformed by new inputs and by the passage of time" ("Computational Mechanics" 2). Patterns are evident when a causal link develops, despite entropy, to produce an outcome. But the link is only probable, never certain or deterministic, because "it relies on exact predictors that are 'fuzzed up' by noise" (5). For Shalizi (SF), information is "just something about probabilities."

As Johnson writes in *Fire in the Mind*, meaningful events that exhibit information content are rare and surprising in a world that is dominated by seeming randomness and noise. For example, if you hike through the mountains and stumble upon an arrowhead, you are excited because that ancient tool contains more information than a rough piece of rock. It is also a rare and surprising find (122). Yet, as the Fontana and Buss model of molecular functions reveals, every bit of matter contains some information that has directed its formation.

METAPHOR HARMONICS: WHO INTENDS THE BEE TO BE YELLOW?

A key difference between Fontana's metaphor of a molecule as a function and Shalizi's metaphor of causality as a pattern is that the former would seem less subject to multiple associations than the latter. Upon hearing the phrase "mathematical function" or certainly, "Church's lambda calculus," different readers are not likely to draw idiosyncratic mental pictures. The word "pattern," however, does invoke images of clothing, art, human behavior, and so forth. So while it seems that Fontana and Buss have fixed their metaphor as much as can be done, Shalizi and Crutchfield and other computational physicists have less control over the cognition that arises from their term.

But aside from offering readers a degree of familiarity and comfort by presenting a technical concept in a colloquial manner, are there enough benefits to metaphor to justify the risk of spurious associations, undesirable harmonics? Perhaps the computational physicists would be better off with a neologism. Clearly in some cases unintended associations can bring on new insights, as we saw when the metaphor of light as a "wave" spawned an entire new understanding of the tentative, probabilistic nature of subatomic particles. The flip side is that some unintended associations can distort a theory, especially in the popular media. Perhaps, as the economist Solow fears, harmonics can even turn a theory into mush. In this section we will consider what the SFI scientists have to say about the possibilities and perils of metaphor harmonics, and look specifically at some SFI terms that seem to resonate with multiple meanings.

We do not have to look far for examples. Shalizi (SF) notes that in models of causality that assume the present retains bits of information about the past, there is the concomitant assumption that some information is lost to entropy. So each causal state is threatened by the chance that linking information will not be communicated. Pioneering information theorists captured this risk in the term "uncertainty." But uncertainty carries with it human associations; uncertainty seems subjective, Shalizi says, as if it "would vary from person to person." In their journal submission, Shalizi and Crutchfield interject a parenthetical solution to the harmonic problems inherent in the old term: "Those leery of any subjective component in notions like 'uncertainty' may read 'effective variability' in its place," Shalizi and Crutchfield write (7). Economists like Arrow, by contrast, use the term "uncertainty" freely because they are dealing with human behavior.

For the same reason, Shalizi is wary of the term "self organization" to describe how states of lower entropy and greater order arise among interacting agents of all kinds—molecules, organic cells, people trading in the marketplace. Chemists refer to this process as "self assembly," which Shalizi sees as having fewer spurious harmonics and, therefore, greater representative value when referring to inanimate objects. "Organization" carries with it overtones of human society, while "assembly" sounds purely materialistic. But the term "self assembly" has its own overtones of engineering design; it is too specific to be applicable to a lot of Santa Fe Institute research, Shalizi says.

Whenever scientists use figurative language, they run the risk that the image it evokes in the minds of an audience may be different from what they intended, especially if the audience contains people of various disciplines. Dulle (SF) has a computer science background, and says that the term "epsilon machine" is troubling to her because the term implies something mechanical. She once asked Crutchfield, "Is an epsilon machine software, is it hardware, or is it none of the above? Is it a mental construct?" Crutchfield suggested that the notion of a "mental construct" was probably accurate. "I'm sitting thinking about the wiring diagram that he must have done to make this machine work because he's talking about inputs, processes—and I'm thinking computers," Dulle continues. "And after the fact he's telling me it's a mental construct. Why didn't he call it an 'epsilon construct'?"

Ostrom's (SF) use of the word "rules" is a clear example of the problem of different images emerging from one word. As a political scientist, she was using "rules" to define an enforceable description of actions that are permitted and prohibited in a social group. Her model dealt with allocation of natural resources, which she considers from an economic perspective as a problem of group decisionmaking. Yet, even those taking an economist's perspective in the audience had reason to question her assumption. "Why begin with rules when much of the economy is alegal and not regulated?" someone asked, suggesting that the term "rules" be replaced with the term "strategies." As Ostrom (SF) says in her interview, agent-based modeling (which is done often at the Institute) uses the term "rules" as "any kind of observed regularity that you can program." On a biological level, enforceability is not a component of "rules," as it would be on a political level.

Obviously, much of the impasse resulted because scholars of different disciplines were all trying to make use of a word that is loaded with harmonic implications. Perhaps Ostrom could have resolved the legal connotations of "rules" if she had avoided the term and simply made the case that people allocating resources make decisions on how to do so. But in opting for the perhaps less loaded term "decisions," she would have been denying an integral component of political science—that groups regulate and enforce behavior. Farmers allocating water rights, or even animals in a pack determining who will be the leader, would seem to be different from plant cells that pursue a "strategy" of gathering sunlight, which collectively causes the entire plant to lean towards the light. Haraway notes that such involuntary actions, referred to by biologists as "tropisms," are obligatory (21). Plants do not develop rules for such action, in the political or animal behavior sense, and yet in another sense, perhaps they do. No governing body enforces the rule that a plant must lean toward the light, but that plant will have less likelihood of survival if it does not. This certainly is a strong consequence. A biologist, then, could argue that he is using the term "rules" in exactly the same way Ostrom is; in her case a political body enforces the rules, while in his case it is nature—the environment—that punishes a less fit plant.

So the definition problem across disciplines for "rules" might seem resolvable by replacing the anthropomorphic notion of "enforcement" with that of "consequence," until we probe further to realize that the semantics dilemma does not have as much to do with the outcome of action (whether enforced or biologically determined), as with the agency of the actor. Individual farmers have the power to make individual decisions and face consequences (a fine, social ostracism, etc.), while individual cells in the plant cannot "decide" anything. Yet, if we look at any plant we can see that some leaves, or even parts of leaves, appear to be healthier than others. Clearly some cells are more successful than others; did they somehow follow a strategy that was tantamount to "deciding" or "choosing" their fate by pursuing a course of action different from other cells?

For Ostrom (SF) and other political scientists, a concept of "rules" requires agency, where the actor behaves with "some sense of normative content." Following Ostrom's definition, we are close to making the claim that "rules" require the consciousness present in human beings or other higher animals. Yet, SFI scientists attempt to dethrone the agent, preferring to see "rules" as patterns of behavior that lead to organization, but that do not require life—let alone consciousness. As we search for a definition of "rules" that works across disciplines we find ourselves in the position of someone laying carpet who smoothes out the carpet in one area of the floor, only to find that this action has created a bump or misfit somewhere else.

This problem of choosing a term that fits all sciences is similar to a problem that physicists encounter when postulating physical laws for how atoms align in a magnetic field. Moore (SF) notes that physicists use the word "frustrated" to describe the phenomenon, implying an intentionality among the atoms that is thwarted: "I fix this problem and I create another problem over here," Moore says. "So satisfying all the constraints is not easy, just as packing a suitcase is not always easy. You figure out how to fit this thing in there, but now this other thing won't fit." In computer science the same problem is known metaphorically as the problem of "inpacking" or "knapsacking," Moore adds.

In trying to smooth out an all-purpose definition of "rules," we have created many troubling bumps. Upon closer look it is clear that these bumps are part of the larger problem of intentionality, a form of causality, which we encountered earlier as one of the central metaphysical problems for science. Do farmers allocating water rights act according to intentions that cause a certain outcome? It seems obvious that they do so. Do plant cells act according to intentions that cause them to thrive or perish? The answer is not as clear, and begs all sorts of questions about the role of will and consciousness in intentionality. Philosophers for centuries have debated whether human beings have free will, or whether such apparent free will is just disguised determinism (Jeans 214). For example, individuals at a baseball stadium from close range appear to act according to individual intentions, but from a blimp flying above, the whole scene appears more like a single organism. SFI research inverts the question of will versus determinism by leaving open the possibility that the apparent determinism that drives dumb matter or lower life forms may be disguised agency. Plant cells or electrons may follow processes that mimic decisionmaking.

The question of will in plant cells or electrons would have seemed absurd to Newtonian scientists, who saw all dead matter and much living matter below the mammalian level as behaving according to rote determinism. When quantum physics replaced deterministic rules with probability distributions, however, we were forced to confront the implication either that particles have some "choice" in how they act, or that we, as observers, have that choice. Either way, it becomes more difficult under postmodern science to dismiss the presence of choice and intentionality among agents.

Semantic snags related to intentionality are present not only when crossing scientific disciplines, but also within specific disciplines. Theoretical biologist Michael Lachmann (SF) in his interview calls attention to the term "signaling" as an example of a metaphor that is rich, but also defies precise definition among biologists. Lachmann's research at the SFI focuses on issues of animal communication, especially on the "cost of signaling." A textbook example shows that a stinkbug attracts its mate by giving off a pheromone that also attracts parasitic maggots; the maggots are the "cost" of a successful signaling strategy (Bradbury and Vehrencamp 8). Lachmann says his research often gets tangled in trying to define "signaling" in relation to intentionality. "So if we see a bee that is yellow, then it might signal that it is dangerous to eat." Lachmann says. "But does that mean then that some beetle that is brown is signaling that it is not dangerous to eat?" Perhaps a brown beetle evolved according to a strategy that is based more upon camouflage, which is a concept semantically opposed to signaling. Signaling implies calling attention to oneself, while camouflage carries the connotation of hiding. The possibility of camouflage as an evolutionary strategy is suppressed, however, by the signaling metaphor, unless we stretch the metaphor to allow that a brown beetle could be "signaling" that it is not there, that it is just part of the background. The metaphor of signaling would function in the manner that Boyd and Pylyshyn described in the metaphor theory chapter, whereby the imprecision of a metaphor and its harmonics, when compared to observed reality, serves as a prompt for scientists to modify their theories.

Signaling not only implies calling attention to oneself, but it also suggests a purpose that is realized by some living agent—another harmonic association that is potentially misleading. An insect's color is purely the product of evolution. Can we use a metaphor like signaling, with all of its connotations of agent-directed purpose, to refer to a DNA-coded process? "So here is the problem of intentionality, like who intends the bee to be yellow?" Lachmann asks rhetorically. Science is replete with terms that imply will and intentionality and these connotations persist or are given new meaning as our understanding of reality changes. Foucault and other knowledge theorists have shown that scientific phenomena for the ancients were seen to resemble human narratives; for example, a drama of deities played out across the sky each night in the form of rotating constellations. Foucault says, "The world is covered with signs that must be deciphered." . . . (32).

At this level of deciphering, humans made meaning by surmising resemblances between their stories and those of the physical world; hence, intentionality in a human sense was implied at every level. Moore points out in his interview (SF) that in Classical physics, a "rock literally wants to fall" to the earth.

Most objects in pre-Classical science were imbued with some degree of intentionality until new atomistic theories dissembled reality into component parts and implied that will inhered only in the collection of those parts, not in the separate atoms. "Never suppose the atoms had a plan, nor with wise intelligence imposed an order on themselves. . . ." Lucretius argues, claiming that order happened only after chance organization of those atoms (172). From the contemplations of early atomists until the present day, scientists have assumed that intentionality is a higher-level activity that cannot be ascribed to brute matter. Yet, the language of intentionality persists at all levels; we still say, for example, that "nature abhors a vacuum," as if all of nature itself had a will. Conversely, at the atomic level, we say that electrons "repel" each other, but they are "attracted" to protons, implying an anthropomorphic response among particles. As Moore notes, scientists talk about "frustrated iron atoms" or cellular automata that "vote." Physicists referring to phase transitions among groups of atoms speak of "a compromise between trying to reduce energy and trying to increase entropy," he adds.

Dulle, the Santa Fe Institute's director for business relations, says in her interview that the anthropomorphisms of science are particularly troublesome for a non-scientific audience attempting to unravel SFI theories. She once kept of list of terms used by scientists that seemed to ascribe human will and emotion to inanimate objects or computer simulations. For example, a term like "the robustness of a system," often is used to suggest that a system—such as a computer model of unspecified agents or "cells"—is surviving and moving to greater levels of fitness. "Immediately when I heard those terms, it took me right out and flipped me back into social systems thinking, you know, social interactions," Dulle (SF) says. She wonders how scientists can use such anthropomorphic terms for non-living systems. One SFI scientist acknowledged the problem of human associations in his lecture, Dulle recalls; this helped the audience avoid confusion. He made it explicit that he was not talking about living entities when he used the term "agent," even though his language suggested he was. "Now you're traditionally probably thinking of the guys in the trench coats and the hats when you hear the word agents," Dulle says, quoting the scientist. "But that's not what I mean here."

Johnson (SF) recalls writing a science journalism article about slime molds sending out signals. The implication is that signaling is a form of "communication," a term also fraught with semantic problems. Does communication require consciousness, or can an unconscious mold communicate? We are back to the yellow bee problem. "I did get an email from someone complaining about that thing, that I was ascribing intentionality to these single-celled creatures," Johnson says.

Clearly the problem of intentionality is a boundary problem that is particularly troubling in biology, which at its foundations is a science of classifications. The difference between two species, for example, may seem trivial for biologists looking at a collection of organisms at a given point in time. If two organisms cannot mate, they are of different species. Yet for Darwin, and for paleontologists looking at life across time, it is not as clear when new variations become new species. The distinction between non-life and life can be equally as opaque. Consider a virus, for example, which is just a section of code that cannot metabolize or reproduce outside of a host cell, but which can "exist" in an inert form until it encounters such a suitable host. While the science may not define a virus as a true living agent, clearly the language used even among virologists implies intentionality. Robin Marantz Henig, a science writer, notes that viruses elicit many colorful anthropomorphic metaphors, such as "pirates of the cell," "submicroscopic hijackers," or even "teenagers run amok"— the latter image implying that viruses were once pieces of our own human genome that broke off thousands of years ago (58–59).

Cowan (SF) argues that a living organism is one that exists long enough to replicate. In this sense, a virus as code that reprograms an organism to facilitate its spread could be seen as living. As rhetorician of science Richard Doyle argues, modern biology has driven away the metaphysical notion of "life" as some kind of vitalism; the spirit of life has been supplanted by an algorithmic process similar to the virus, which is buried in the genome (13). Certainly an algorithm cannot have a will; it iterates blindly. Such an understanding of life, then, would invalidate metaphoric harmonics that imply intentionality—or metaphysical life, for that matter—at any level of being.

The problem of intentionality would seem resolvable if we were able to agree on a clear definition of a living organism that did not require a metaphysical presence, and to determine when that organism acted on its own volition. A rock falling, then, would not have intentionality because it is not alive. A bee would not have intentionality in its coloration because it took no action to affect that color. Intentionality implies consciousness, whereby an organism acts with a purpose that requires knowledge of the consequences of such an act. Yet, semantic problems persist. Kauffman (SF) notes, for example, that we are willing to say that a bacterium swimming upstream in glucose "is acting on its own behalf in an environment." Certainly it is alive, in that it moves as a result of its own internal metabolism. But, Kauffman adds, "It's just a bunch of molecules." Lachmann (SF) suggests that avoiding anthropomorphic terms for non-human entities could ease problems of implied intentionality. So biologists might say that cells "evolve into" certain states rather than that they "choose" those states, Lachmann says.

For Kauffman, this semantic dilemma is valuable, however, because it leads scientists to recognize that our anthropomorphic assumptions about what constitutes an autonomous agent may be an impediment to our understanding. Perhaps we are obsessing over the essence of reality to the point that we cannot clearly see its behavior. Kauffman suggests that we can consider a physical system to be an autonomous agent if it is "a self-reproducing system that can do at least one thermodynamic work cycle," in other words, if it replicates and converts energy from one form to another. Such a solution would skirt questions of what those agents are (e.g., living, conscious, and willful) and focus on what they do.

Kauffman's research on self-organization uses the model of a group of light bulbs that switch on or off depending on the actions of neighboring bulbs, creating patterns of apparent organization that have nothing to do with life, consciousness, or will (*At Home* 74–92). He and other SFI researchers offer a kind of phenomenal or pragmatic approach to the problem of recondite definitions by suggesting that we look beyond the definitions of systems to their phenomena, their behavior, their patterns. Phenomenalism is a minor school of philosophy that attempts to avoid the problem of intentionality and causality by focusing on behavior and experience as the source of perceptions, which are the source of reality. Kauffman's definition of an autonomous agent seems to suggest that researchers search for metaphors of action rather than metaphors of being.

Casti (SF) offers a similar approach toward arriving at a definition of "rules;" his definition does not conjure up the idea of consciousness or other philosophical traps. Look at a rule simply as a process of transformation, Casti suggests. "The mechanism that actually transforms the input into the output—that is the rule," he says. Yet, Casti recognizes that such a catchall definition ignores interesting scientific questions about what happens inside the minds of members of a rule-based society, or during the evolutionary process. The action that leads to the transformation is treated like an unknowable event inside of a black box—perhaps an honest phenomenal approach, but not very intellectually fulfilling.

Lachmann also has considered taking a phenomenalist's approach to problems of word meaning. He recognizes that it is possible to use the term "signaling" to describe a variety of specific biological phenomena without defining the term in the abstract. "I think this is what should happen very often, but I think very often it doesn't," Lachmann (SF) says. But try as he might, Lachmann cannot avoid being led to abstractions. Whenever he uses a term like "signaling," he is confronted with its paradoxes. Perhaps the only way to try to resolve these is by coming up with a definition that is abstract enough to embrace all the metaphoric harmonics.

Even if we ignored the abstract essence of a term like signaling, or the essence of an autonomous agent, and focused solely on actions manifest in the world, questions of intentionality would resurface by focusing attention on what constitutes autonomy. When Kauffman develops models of light bulbs in a circuit that change colors according to interactions with other bulbs, is each bulb autonomous, or is the system itself autonomous? We become mired in semantic confusion when the behavior of individual agents affects the success of a whole group, which then affects the behavior of other individuals. Metaphors like "signaling" that imply some kind of intentionality cannot distinguish between individual goals and group goals, as Lachmann's simulations shows. At the Institute he simulates the behavior of an ant colony using an IBM laptop computer. The "ants" are in various states of wellbeing (shown in the model as different colors). An ant's state changes according to how it interacts with other ants, with the implication that some degree of interaction is necessary for the success of each individual ant. So, Lachmann (SF) says, "You could even ask what's the intention of the colony. Does the whole colony have an intention?" That is, is the colony signaling that a degree of collective behavior is good?

Lachmann's questions about intentionality among a group of computer ants or among individual members of that group bring into focus another semantic problem that particularly haunts the biological sciences: What do we mean by the term "individual"? As Keller shows in her research, much of the debate in evolutionary biology since Darwin has reached an impasse over the question of where natural selection occurs. "For Darwin, the object of evolutionary discourse, the biological individual, was unambiguously the organism," Keller writes. "But since Darwin, the locus of biological individuality has become more ambiguous" (145). As Keller writes, a post-Darwinian individual—that is, the unit of life or lifelike agency that struggles for resources—could be a species; a group within a species; an organism, i.e., a single member of the species; or the genes of a single member. For Keller, how a scientist defines an individual is related to how much value he places on group activity versus that of a single agent; this is a question of "methodological individualism versus holism" (147). Evolutionary geneticists such as Richard Dawkins, who reduce evolutionary significance solely to the level of what he calls "the selfish gene," have promulgated an almost corrosive view of the value of society, or even the person, Keller suggests. The concept of "individual" in Dawkins' theories is reduced to that of genetic code, with obvious implications for what it might mean to be human.

We might look to other Institute metaphors possibly to restore the seat of individuality to a higher life form than mere bits of genetic code. Fontana, the theoretical chemist involved with lambda calculus, has taken the common word "neutrality" and used it to argue that the genetic code underlying an organic compound such as RNA (ribonucleic acid) can change frequently without affecting the performance of the compound until at some point the compound seems to "suddenly" evolve to a higher level of performance. Such "neutral" changes in the underlying genetic sequence make evolution possible, Fontana (SF) says, because they allow changes to occur without risk of wiping out the system. To explain the concepts of neutrality and performance, Fontana draws an analogy with an automobile engine whose performance depends on many parameters. If you want to improve the car's performance, you need to change parameter settings, but many parameter settings yield the same behavior, Fontana says. "If there was no neutrality you would be stuck at the moment because you suppose that all neighboring parameter configurations are worse," he says. "Then you're left guessing. You can guess up till the cows come home and you're unlikely to find the right parameter settings. . . ." With neutrality, you can change the settings gradually without damaging the engine until you arrive by chance at a point where the next change causes the performance to go up.

An implication, or harmonic, of this metaphor is that a car engine's performance is more than a single wire. Life, it follows from the analogy, is something beyond the reaches of an individual bit of code. Otherwise, a change affecting an individual gene could never be deemed "neutral." A human's life cannot be the summation of the "lives" of each gene in his body even for a strict adherent to the selfish gene reductionism. Still, Fontana acknowledges that the neutrality may only be temporary; eventually the compound evolves. So it seems appropriate to ask if a change can be called "neutral" when it has no apparent immediate effect on a compound or an organism, yet over time, in combination with other changes, it leads to a profound impact—an optimized performance.

A concept that is related to performance is that of "fitness," which in biology describes the success of an organism in being able to reproduce fruitfully. "Performance" in Fontana's model describes the successful behavior of an individual, a phenotype, while "fitness" is related to the successful proliferation of the underlying genetic code, the genotype. The aggregation of fitness possibilities that different genotypes can reach through evolution is known by a metaphor we have already encountered, "fitness landscape." This metaphor dominates biological models at the Institute, calling to mind the image of a rugged terrain with peaks and valleys. In some versions of the model the deepest points are the fittest, while in others it is the peaks that are fittest. The metaphoric image is rich with theory-constituting harmonics, or images. For instance, we can picture a genotype as a climber who reaches a high peak only to find himself surrounded by even higher peaks that he cannot reach without first climbing down from his local hill. Some genotypes may wind up in such a position, where they cannot reach a condition of improved fitness without first incurring setbacks. Perhaps fish that evolved into amphibians had to spend generations wallowing and dying in mud—an obvious setback—before they evolved the legs necessary for locomotion.

Kauffman and other researchers struggle with the implications carried by this metaphor, however, because the concept of a landscape is so evocative of human-centered associations. Real landscapes are fairly permanent terrains that can be visited by living creatures, but in a genotype fitness landscape, it is the creatures themselves that make up the landscape. As the physicist Moore notes (SF), the landscape metaphor is useful, but these landscapes are always changing. "There are probably situations in biology and in economics where the landscape is relatively fixed, at least for awhile," Moore says. "During that time you can think of it as climbing toward a peak, but then there are other times when it is changing too swiftly for that metaphor to be useful."

Even the concept of "fitness" on its own is problematic because, by definition, it stands only for the reproductive success of a genotype, which depends upon a whole organism. Measurable factors such as an organism's intelligence or how long it lives, or even qualitative factors such as the success of an organism's life, are not taken into account. Yet the implications of "fitness" suggest that those things should matter. Otherwise, in theory, a cockroach could be more fit than a human being, given its reproductive success. Fontana (SF) acknowledges that some definitions of fitness now take into account the life history of the organism that carries the genotype. Perhaps the metaphoric harmonics that radiated from the term "fitness" propelled biologists to develop more complex definitions, and to confront the unsettling implications of a life's value being reduced to the transmission of code. The harmonics would not sit still.

In retrospect it becomes clear that scientific metaphors that ascribe neutrality, agency, individuality, autonomy, and intentionality to the products of evolution cannot perform their role of representing knowledge without also abruptly calling that knowledge into question. When terms such as "signaling," "camouflage," "rules," "fitness," "fitness landscapes," "self-organizing systems," and "function" operate metaphorically, they do so by drawing associations that ultimately penetrate to the fundamental questions of science philosophy—what is life, free will (or agency), individuality, success, and so on? Perhaps it is the relentless demand for introspection that makes such metaphors valuable as tools of cognition. Metaphor is a trope that is constantly in motion, forcing its users to repeatedly confront its contradictions. It is becoming clear as we follow the comments of these SFI scientists that it is the contradictions inherent in metaphor, the dissociations, and the demand for accommodation that make metaphor an essential device of representation. True cognition follows when the representation dances with its own paradoxes.

METAPHOR HARMONICS: EMERGENCE, THE BRAIN, AND NEURAL NETWORKS

The unease a scientist may feel in confronting the contradictions of metaphor is found not just where the boundaries are fuzzy between non-life and life or involuntary behavior and intentionality, but also at higher levels of order, where the distinctions would seem to be clear. Consider the seemingly non-controversial assertion among many SFI researchers that complex behavior and order arises from the interactions among various agents. When a swarm of bees, or colonies of ants, or pixels on a computer screen achieve success that depends upon members of the group relating to each other, a SFI scientist would likely proclaim that a complex order has "emerged" from the group. This is similar to the organicist perspective in biology that Haraway describes, where parts of a system interact at all stages of its development. Young tissue in a developing salamander fetus, for example, functions "as a unit" to allow the whole creature to emerge (*Crystals* 158). Order, then, is the result of interactions among the different parts of an organism at all stages of development, not just when the creature is finished.

Yet, as Arrow (SF) notes, the concept of "emergence" carries with it associations of unpredictability. In other words, there should be no way to know the outcome in advance without letting the agents first interact. "I've not really seen a clear-cut definition of emergence," Arrow says. "It always has . . . the idea that somehow it has a psychological element in it. It's a surprise. But if you have seen it happen a couple of dozen times, it is not a surprise." Arrow's observation points to a potential limitation of the emergence metaphor. The metaphor suggests room for variations of outcome, which implies that the agents have some control over a course of events that cannot be predicted. But if the same outcome obtains in all cases, or even in most cases, the metaphysical implications of emergence are lost. If the tissue almost always grows a fit salamander, can we still say that such an organism has emerged from the interacting components? There are no surprises in a deterministic system, although a metaphysical term such as "emergence" can mask the determinism.

Associations between inanimate deterministic materiality and indeterminate living systems also are present whenever electrical switching metaphors are used to explain what goes on inside the human brain. The metaphors borrowed from the physical sciences have been used to reduce complex biological and social phenomena to mechanical operations. Hely (SF) notes that the process of human thought has been explained by metaphors taken from whichever physical science is dominant at the time. The brain as a machine of Newtonian physics gave way to the brain as a computer, Hely says. As we have seen in metaphor theory, however, the metaphor can flow both ways; computer scientists in the mid-twentieth century came to view computers as biological brains at the same time that biologists came to view the brain as a computer. Both were modeled as a "neural network," Hely says, which is a system of nodes and connectors that interact to process information.

Neural network models can be a physical collection of wires and switches or, in the SFI case, computer simulations of such systems. At first these models were useful both in biology and in artificial computation sciences to depict a type of hub and spoke operation in which sensory information (input) is routed to various parts of the system. The system then responds with thoughts (output). An artificial neural network is a computer that is modeled after the human brain, where nodes, or "neurons," are assigned certain values. They exchange signals and gain or lose value to the system according to how many exchanges they participate in. Hence, neurocomputers are able to achieve proficiency in whatever task they do (robotics, military surveillance, etc.) by "learning" the task—a process that depends upon those signal exchanges.

But the metaphoric association drawing parallels between the brain and a computer has not proven strong enough to withstand scientific scrutiny. The harmonics ring out with contradictions ("dissociations," to use the Perelman Olbrechts-Tyteca terminology) that cannot be resolved without modifying the metaphors that form the foundation for the model. Hely says that biology and artificial computational sciences have split so that the two no longer assume that the same model of computation, which draws on the same metaphors, applies for both natural and artificial thinking systems.

Hely describes how the brain-as-computer, computer-as-brain metaphor has been dissociated. In biological and artificial networks there are learning rules, he says, whereby knowledge is seen to inhere in the connections between cells in the system. The most common rule, known as "Hebbian learning," is expressed in the form of a common word play trope. It declares "cells that fire together, wire together." In other words, if one brain cell, or a cell in an artificial network, fires an impulse causing another cell down the line to fire, then those cells together have a strong connection that could be seen as a cognition. We know, perhaps, that the sky is often blue because brain cells that hold the image of blueness and the idea of sky usually fire together. Likewise, a military intelligence neural network hooked up to a camera could detect the presence of a tank in a landscape by picking out vague images, which the computer associates with tanks.

In artificial networks, however, the programmer could "train up" the network by assigning a numerical value to the relationship between cells and then letting that number increase (or "strengthen" the association) every time they fire together. But in biological networks, Hely says, it is difficult to reduce the process to a numerical strengthening because it is not clear exactly how the neural connections work and what steps are involved. "It's a much messier process . . .," he says. Clearly the brain is too complex for the brain-as-computer metaphor to work. Once we listen to the harmonics emanating from the metaphor, we realize that those harmonics do not resonate with each other. "So the split was probably caused primarily by biologists and some of the computational neuroscientists saying that artificial neuron networks are very efficient for solving certain tasks, but that they're no longer giving us the insight into what's going on in the brain because they're just oversimplified," Hely says.

Both the brain and artificial neural networks have been represented by metaphors built from electrical engineering, where signals move through a circuit. So in explaining brain physiology to a lay audience, Hely (SF) might refer to a nerve fiber not by its scientific name, "axon," but simply as a "wire." A synapse becomes a "connection." But circuitry metaphors are not always precise enough to accurately represent new understandings in biology. Hely offers the following example: An electro-mechanical connection in the brain, known as a "synapse," implies a kind of "plug and socket," but the structure of the synapse also involves cellular chemistry that would not be present in an electrical circuit. "And so it's much more complicated than perhaps just putting a socket together," Hely says.

Hely, however, does not dismiss the plug-and-socket association even though it is an imprecise representation of the brain's physiology. In some cases even an inexact metaphor is useful for theory building. "So I guess you're losing some specificity, hopefully gaining clarity . . .," he says. If all you are interested in is the movement of a signal along an axon then you may not care about the buildup of chemical imbalances at the synapse that precede the electrical spike. Similarly, complexity scientists may not care about the source of a signal at one site so long as it reaches a threshold level and flows to another site. John Holland, a computer scientist affiliated with the Institute, notes in his 1995 book about complexity that all complex adaptive systems involve nodes or agents that exchange information; this flow of information back and forth among agents can give rise to new agents and new connections that allow the network to adapt to its environment (23). Human nerve systems, the world's ecosystem, and the Internet all function according to this pattern of information exchange among agents, Holland says. This recalls Bak's argument that a certain amount of generality is necessary when a science attempts to describe commonality across various real systems.

Yet, when we find that the same metaphors do not apply iso-morphically across biological systems and artificial neural networks, we are assured once again that the power of metaphor is in revealing the differences between apparently similar systems or phenomena. In another example of how metaphors reveal differences in fields, Hely explains how information is passed in the brain when nerve cells send and receive messages at a greater frequency than what occurs during normal ambient brain activity, which he refers to as "housekeeping." So a person daydreaming who suddenly sees someone she knows may experience a burst of neural impulses as the sensory image links up with the brain's memory of that person. Hely's notion of "housekeeping" suggests that the brain has degrees of cognition—some ambient, others more powerful. The ability of the brain to both keep the heart beating, a "housekeeping" task, while walking and reciting poetry, is what distinguishes human thinking from machine "thinking." A neuron in the brain has as many as 100,000 synapses, while artificial neural networks have less than ten such connections ("Neural Networks," *Encyclopedia* 930). Clearly the brain is capable of much more subtle information processing.

Biologists describing the physical behavior of the brain have drawn on a collection of metaphors from various domains, not just the domain of computer sciences. Metaphors are freely mixed, sug-gesting that even when scientists select various metaphors to build a model, they do not feel restricted to the same domain or the need to appear consistent. In describing the brain's activity, Hely uses anthro-pomorphic metaphors (cells become "excited" or "shout louder"), military metaphors (a cell "fires"), and metaphors from electrical engineering mixed with geography (cell "spikes" are felt "downstream"). Moore (SF) points out that some agent-based models at the Institute freely mix metaphors from biology and economics. These models refer to agents "eating" or engaging in "combat," while also referring to the effort necessary to maintain the agent's metabolism as a "tax." As Hely explains, "Metaphor is used in trying to explain concepts badly understood or overly complex. If you're trying to come up with a model or think about what is going on then you're trying to use the best available metaphors to give you a handle on that."

Metaphors as the seeds of scientific models functioned for brain biologists and artificial computation scientists perhaps as temporary scaffolding, to use the concept that Shalizi (SF) introduced. Yet it seems clear that without the tension between signifier and signified that metaphors generated, many valuable insights about the nature of thought at the natural and artificial levels would not have surfaced. Once again it becomes clear: metaphor is theory-constitutive not in spite of the harmonics, but because of the harmonics. As we remember from Chapter 3, Boyd specifically cited the metaphor of the computer as a brain being "theory constitutive" because it induces scientists to unravel each aspect of the theory contained in the metaphor. Metaphors elicit associations that force the researcher to confront contradictions in his model, a process that Boyd and Pylyshyn described by using Piaget's concept of "accommodation."

Hely seems to have followed Kuhn's implicit advice about accommodating metaphors, which holds that scientists should mix and modify metaphors to best represent reality as it presents itself. Haraway argues that we can detect a Kuhnian paradigm shift in science when metaphors start to converge, in other words, when a branch of science begins filling up with metaphors from a specific domain—computer metaphors for the brain, for instance (*Crystals* 196). An equally clear indicator of when the paradigm is dying out and making way for a paradigm shift, it would seem, is when metaphors begin to diverge, as has happened subsequently in brain biology and artificial computation sciences.

Biology may be diverging from the electrical-mechanical foundations of computational sciences, but at the same time, computer science is moving away from those foundations toward a more biological model of computing. Computer scientists using the word "virus" to describe malfunctioning (even malicious) computer code are clearly envisioning computing as something more organic than the mere exchange of electrical impulses. Computer code itself is akin to the biological code of life—DNA, deoxyribonucleic acid. Computer scientists affiliated with the Institute have pioneered research on how to make computers function as if they had a biological immune system to fight off viruses. The lesson for scientists using metaphor (and for rhetoricians and technical writers) is clear: once a term like "virus" is used metaphorically across fields it becomes impossible to ignore the deductive logic that comes with the metaphoric associations. If a computer can have a virus, it follows that the computer must be more organic and less mechanical than originally thought.

Hence, the association between the brains of living organisms and computers remains useful, but in a modified version that draws attention not to the brute electrical process, but to the subtle results of that process—an exchange of information that can be either useful or harmful to the system. Both biology and computer science are approaching the field of quantum physics, where information has replaced materiality, where particles are seen as temporary states organized according to the erratic grammar of probability. Moore (SF), sees the link between material states of being and information exchange in the concept of entropy, a process whereby ordered material states gradually lapse into states of disorder despite the various "laws" of physics that constrain their behavior. Hence, physicists can speak of events at the subatomic level (in which particles decay or transform into other particles) not so much as a change of state of being, but as a change in the amount of information contained in the particles. A particle that decays is one that loses information. "Right now there is a very hot field of quantum computation, trying to understand whether quantum mechanical systems could do computations in fundamentally different ways than classical systems," Moore says. He is seeing the interaction of particles as a computing process, which recalls the Fontana and Buss metaphor of molecules as semantic units. The metaphors of science clearly are converging to a new understanding of biology, chemistry, computer science, and physics as sciences not of matter, but of meaning.

EQUILIBRIUM AND THE PRISONER'S DILEMMA

Metaphoric harmonics that radiate from terms used to develop theories of brain physiology, artificial intelligence, bee swarms, causality in physics, and similar abstract questions of science would seem without question to be valuable tools of cognition that spawn new insights. If metaphor harmonics generally are valuable for abstract theorists, we are then led to ask if they also are valuable when those theories are turned into applications. Perhaps those harmonics become dangerous when they escape the corridors of science academe. In this section it is worth looking at the policy implications of metaphor harmonics that are used, but were not necessarily developed, by Santa Fe Institute economists. This study should shed some light on the question of what happens to science metaphors and their harmonics when they get loose in the outside world.

Economics as a social science studies how the preferences of individuals, or of individual firms, collectively lead to the phenomenon of the marketplace. Standard neo-classical economic theory that first arose in the Enlightenment posited a view of the world that was gleaned from Newtonian physics, which suggested that the economy was rational and deterministic, a series of interacting forces that could be measured. Consumers and producers acting in their own best interest would drive the economy to a state of "equilibrium," a condition in which prices settled to a level that could not be moved by the action of any individual, where no individual's economic wellbeing could be improved without hurting another's. Much of the research in early twentieth century economics focused on whether the economy, beset by a terrible depression, would ever return to an equilibrium condition.

The word "equilibrium" in economics is generally taken to be a metaphor borrowed from physics. The term derives from two Latin words that mean "equal balance," suggesting a state of being in which physical weights cancel out each other's forces. Physical weights are real in physics, but only metaphoric in economics. Economists for more than a century have built theory around the term "equilibrium," but not without occasional moments of self-doubt—wondering if their semantic effort to imbue legitimacy to an inexact social science is merely a case of "physics envy."

The problem with the "equilibrium" metaphor for those who resent the envy is that it may imply a certain and determined state of wellbeing in the world, a status quo, a settling of the economy at its natural level. One entry for "equilibrium" found in *The Oxford English Dictionary*, for example, defines it as a "well-balanced condition of mind or feeling" (257). Henry William Spiegel's extensive history of economic theory, *The Growth of Economic Thought*, suggests that the

idea of equilibrium has its roots in the ancient Pythagorean philosophies. Spiegel writes: "From these investigations the notion of harmony was derived, which in turn has an affinity with the concept of
equilibrium that was to occupy a central position in the economic
thought of later generations" (12). So the term "equilibrium" could
have normative associations of goodness and harmony that might be
used to palliate society's sometimes-uneasy truce with a competitive
market economy, one in which it is inevitable that many less successful individuals are hungry and poor.

Arrow, the economist, has focused much of his life's work on
equilibrium theory and its implications for the welfare of society. For
Arrow, "equilibrium" has a precise mathematical meaning in economics; it is not particularly metaphoric and certainly not normative, but
merely a statement of balance in the economy. "If I ask what distribution of weights on a lever will keep the thing in balance, I don't think
of that as being normative," Arrow (SF) says. Those nineteenth century economists who developed equilibrium theory knew its limitations, Arrow says. "Minor writers" attempted to push the analogy
further, he adds, by trying to find economic analogues to temperature, entropy, and other concepts in physics. For Arrow, the state of
equilibrium in the economy is simply the market seeking its own
level. It follows that if society wants to change that outcome to redis-
tribute goods to the poor, such an action would not negate the presence of equilibrium, but would intervene to modify its results.

Cowan (SF) says that physicists and economists recognize that
they are talking about different kinds of equilibrium, determined in
part by time. In physics, for example, a piece of glass may appear
stable, but over a long period of time it reverts back to equilibrium;
that is, it slowly melts. As Cowan says, "In physics equilibrium is
dead; in economics it's alive. We have a metaphor in physics for that
kind of thing, called 'glass,' 'a glassy state,' so the metaphor is 'glass.'
Glass looks like it's stable, in equilibrium, but a thousand years from
now it can revert to sodium chloride and something else. It's unstable
and moving back toward an equilibrium . . . what physics and chemistry would call equilibrium. Economics considers much shorter periods of time in its models." Casti (SF) adds that equilibrium is a
Platonic abstraction. "There are no systems that are in genuine
honest-to-God equilibrium, ever, in the physical sciences," he says.
"Maybe in mathematics you have it. . . ."

Yet, it would seem that while scientists such as Arrow, Cowan, and Casti recognize the limitations of a term like "equilibrium" in economics, policymakers might be swayed by the harmonic associations of naturalness and wellbeing. Again the problem of harmonics may pose little risk to scientific cognition, but perhaps a greater risk to society that attempts to prescribe policy based upon that knowledge. If things are in equilibrium, policymakers might wonder, why not just leave them alone? A whole retinue of metaphors in competitive economics ("the law of supply and demand," "natural rate of unemployment," and so forth) suggests a certain inevitability that would be concomitant with the determinism of equilibrium.

Economics may be in the midst of a paradigm shift as new metaphors from biology are creeping in to replace those of deterministic physics. These metaphors bring harmonic problems of their own, however, not the least of which is the same conclusion that things are as they should be and, therefore, should be left alone. Economics began courting biology as far back as the late 1800s, when economist Alfred Marshall suggested that that models based on evolution of living systems would be more appropriate than those based on mechanics. Santa Fe Institute economists have lent their insights to the new biological model by developing neural networks, feedback systems, learning scenarios, and self-organizing systems to describe how the economy grows and becomes more complex like a living organism. Yet, the issue of time also raises problems for biological metaphors in economics. As Arrow (SF) points out, evolution is unique in that it occurs over tens of thousands of years in which the actual genetic code of species changes. To import this concept into economics and other social sciences that operate on much shorter time horizons requires that major implications of biological evolution be suppressed. Human society and its economy changes over decades, but the underlying genetic code that determines what a human being is does not change over such a short time period. As Arrow says, "I kind of resist the idea that evolution means anything other than something to do with the germ plasm. I'm cautious about that particular metaphor." SFI scientists are aware of the limitations of evolution metaphors in social sciences, Arrow says, "but I think you can get carried away." For Arrow, the evolution metaphor in social sciences is valuable primarily as a "figure of speech" that provides an impetus for new mathematical models, in the same way that inexact physics metaphors in the nineteenth century ushered in useful new equilibrium mathematics.

Seemingly, a new organic paradigm would be less likely than an equilibrium paradigm to spawn Panglossian implications that the world is as it should be. But Moore (SF), a physicist, points out what he sees as equally dangerous implications of the new biological metaphors in economics. Where metaphors borrowed from physics may make the economy seem to be evenly balanced, metaphors from biology can instill false confidence that the economy is more organically natural than it is. "The economy isn't a jungle," Moore (SF) says. In this case, the problem with the metaphor is not that it creates unwanted harmonics, but that it may suppress awareness of the need for an intervening government. One could argue, as Arrow and Ostrom do, that the tendency toward governance emerges in a natural system, and that a good model will explain that emergence. Still, it is not a stretch to see how an emerging-order model of the economy and political system may justify an extreme Libertarian argument that government is an unnatural affectation. SFI resident physicist Murray Gell-Mann has voiced this concern.

A fundamental goal of standard economic policy has been described by the biological term "growth," a term that envisions the economy as an organism that is becoming physically larger, that is, wealthier. Growth in economics often is associated with "progress," which has normative associations of beneficial change. Yet, defining progress in evolving systems is as difficult as defining fitness—perhaps all the more so in human societies, where progress often comes at the cost of lost pristine environments. Adding new real estate developments in the woodlands at the edge of a growing city may be progress in the sense that the economy grows and adds jobs, but not in the sense that virgin land has been sacrificed. Moore (SF) says, "The jury is very much out in biology about whether progress does happen in Darwinian evolution." But the notion of progress is intrinsic to the human community, Moore continues, "so I don't know if it's an idea we can get way from completely, but we certainly need to be more careful about how we use it."

The term "progress" with all of its implications was particularly troubling for nineteenth century scientists as they struggled to write new narratives of change in the physical and natural worlds that would comport with still powerful religious beliefs of the time. If interstellar gases in deep space progressed to become nebulae and individual stars, or if life on earth progressed to become more complex, then the belief that God crafted a perfect universe was called into question. Science historian and philosopher James Secord shows that the concept of "progress" was troubling for scientists even when they were not weighed down with concerns about organized religion. Progress implied that science was a story that moved from beginning to end; this anthropomorphic assumption threatened to align science with inexact literary forms of human knowledge. Secord writes:

> Fears of association with Enlightenment cosmologies tempered enthusiasm for progress as a structuring device, for a narrative backed up by material causation could be read as making humans no more than better beasts. Moreover, a strong narrative line put scientific exposition uncomfortably close to novel writing. (59)

The foundations of economic theory beyond the physics model lie in human psychology and the assumption that people behave rationally most of the time. Economists, sociologists, political scientists, and biologists all assess the impact of rational behavior via the methods of "game theory," first developed by twentieth century mathematician John von Neumann and perfected for economics by John Nash. It is useful to consider game theory as a way of asking whether metaphors that once may have been rich with harmonics can settle down to represent precise scientific meaning, or whether the harmonics persist. In the case of game theory, it is also appropriate to ask whether society needs to keep those harmonics alive in order to mitigate the potential misuse of game theory in public policy.

The original game-theoretic scenario is the "prisoner's dilemma," in which two prisoners accused of conspiring to commit a crime are interrogated separately by police. Each prisoner's fate depends on whether he fibs on the other ("defects") or remains silent ("cooperates"), and on what the other prisoner does. This theory of decisionmaking was used throughout the Cold War by U.S. policymakers to predict the behavior of their Soviet counterparts. Game theory is now a standard tool for understanding decisionmaking among all kinds of agents in science.

Lachmann, the Santa Fe Institute biologist we have met, uses a version of the prisoner's dilemma in his research on signaling among animal populations. He says that the prisoner's dilemma is a model of how populations behave, but that it no longer evokes metaphoric associations of real prisoners or even of real cooperation and defection. The model has been used for so long in science that it has lost any connotative power. So a scientist can use a prisoner's dilemma model even for purely physical phenomena, such as the alignment of atoms in a magnetic field. "Well, any one whose spin is pointing up is cooperating; any one whose spin is pointing down is defecting, or something like that," Lachmann (SF) says. Such a model of atoms in alignment would fit the game theoretic matrix that Lachmann developed for biological agents. "But it still doesn't mean that atoms cooperate," he says. "It just means that what you called cooperation holds here. . . . It doesn't have anything to do with prisoners; it doesn't have anything to do with the original set up."

For Lachmann, the term "cooperation" does not imply human sharing for a common goal. Even cells can cooperate, he says. Game theory for Lachmann is an accepted research method, but not a metaphor; the language is purely representational in that it stands for a method without evoking images associated with that method. So you do not go back to the original game. As Lachmann (SF) says, "It has already been completely abstracted. . . ." The age of a metaphor perhaps determines its volatility; for Lachmann, old metaphors like "prisoner's dilemma" would be settled.

Still, it seems that the age of a metaphor alone would not dampen its ability to inspire fresh and unexpected associations, especially as that metaphor reaches new audiences. Recall that Max Black and other metaphor theorists subscribe to an interactionist view of metaphor, which holds that each word in a metaphoric pair influences the way we perceive the other. So if we are using a prisoner's dilemma image to describe the behavior of atoms in a magnetic field, we see the atoms in a "dilemma" over which way to align (which ascribes intentionality to the atoms). But we also see the human prisoner behaving according to the physical laws of particle alignment. Human decisionmaking, then, is involuntarily rational.

Once again one of the most interesting philosophical issues that results from metaphor use in science (even when the metaphor may have long since solidified into a term of methodology) is that of intentionality in response to information. By being forced to confront a term like "cooperation" and to dismiss its old metaphoric implications of intentionality as no longer relevant, scientists like Lachmann may be uncovering new understandings about reality. Perhaps they are suggesting that "choice" for all agents, even living creatures, can be calculated and almost rational, without being intentional. What constitutes choice for a human prisoner and for its metaphoric equivalent, the atoms in a magnetic field, may simply be nonrandom action. For the prisoner, that action is motivated by information he has about his crime, the likely punishment, and the psychological makeup of his accomplice. For the atom, action is motivated by "information" it detects about a consistent alignment of similar atoms in a magnetic field. Will for the human becomes an afterthought in decisionmaking rather than the prime motivation for such acts. Will could be seen as the involuntary response of an agent to information, a temporary act of ordering reality and restraining entropy.

The potentially disturbing implication for society is that social and political choices may be the result of games played by rational human beings who are responding involuntarily to information. Ironically, a game-theoretic approach to decisionmaking may be more appropriate at lower levels of agency—perhaps atoms in a field or swarms of bees—than at the human level, even though the terminology is heavily anthropomorphic. As Lachmann (SF) notes, agents in game theory only have two choices: cooperate or defect. Yet humans are skilled at coming up with hedge solutions in between. Perhaps it is time for scientists to develop a neologism that removes the connotations of human intentionality from game theory. Just as scientific representations of the human brain and the computer have diverged because of inconsistencies made obvious by the metaphoric harmonics of terms shared across both fields, a similar divergence may be appropriate with respect to disciplines that use game theory.

Doing so would reduce the risk that policymakers might reduce messy human decisionmaking processes to a simple matrix. When game-theoretic methods are used to model the economy of society, as in Ostrom's research, the simplified world of an either-or choice is elided in the model's matrices. By suppressing the harmonics of a term like "cooperation" to allow the model to work for non-living systems, scientists may inadvertently be sending the message that even among humans, the contingencies of life are irrelevant to how we behave. Haraway's observation quoted earlier in this chapter could apply to game theory: the implications of language may be more difficult to hide than the implications of mathematics.

Moore (SF) asks whether decisionmaking matrices such as the prisoner's dilemma are "better than telling the story in English." His question reflects the larger concern he expresses throughout his interview that science reduces reality to elemental units, and then describes events in the world—including human behavior—as determined by those units. Elite policymakers are mesmerized by the methodology, which seems to remove the messy humanness from human behavior.

Newtonian mechanics abolished the idea of free will and replaced it with determinism. Twentieth-century computational physics may have restored will by finding that the end result of a sequence of events in the world cannot be known until all the events have played out, suggesting that reality is subject to the intervention of willful agents. But just when will returned to physics, it was being cast out of economics and biology with metaphors suggesting that game-theory explains all behavior at the phenotype level, while genetic code inscribes a master plan for how that phenotype might develop.

"I don't like the idea of genetic determinism," Moore says. For example, human genes do not have maps of the hand, he says. What genes have, he continues, are instructions that control the timing for when cells should reproduce. "It's an incredibly intricate and beautiful process" that leads to the development of something as complex as a five-fingered hand. Moore is keenly aware how science metaphors affect popular understanding, such as when the notion of a genetic "map" implies that free will is an illusion. "Each of these metaphors is going to have what you may call moral and amoral aspects," Moore says.

We have seen that the challenge of representing scientific knowledge using words is especially compounded when those words are overtly metaphoric, as the first part of this chapter has shown. Yet, metaphor as one of the most important tropes of rhetoric certainly has a theory-constituting role at the Santa Fe Institute and throughout science. Metaphors and mathematical equations work together to produce powerful insights into the nature of reality. But by transporting ideas from one term to another, metaphors conjure up harmonic meanings that force the theorist to confront implications, even paradoxes, of a theory. At the Santa Fe Institute these harmonics reveal philosophical questions underlying science: What constitutes intentionality? What are the differences between a biological brain and a computer brain? What is individuality? Do we mean something different if we say that a molecule exists or if we say that it functions? What is time and how do different spans of time affect the meaning of concepts like equilibrium? How do we ascertain causal links between events? Is information the élan vital that makes matter do things?

The way we resolve these philosophical questions, however tentatively, leads to direct applications in human society. Asking, for example, what constitutes progress prefaces policy applications that would attempt to incubate the seeds of such progress. As we pause at the end of this chapter, we do so with the understanding that metaphor harmonics in science have powerful social implications. They call into question the very essence of humanity.

5
Science Writers Looking for Their Audience

As scholars of the rhetoric of science have shown, the act of presenting knowledge cannot be separated from the act of developing it. This is consistent with modern rhetorical theory, which situates the message among the speaker or writer, his purpose, and the audience. The problems of representing that message, and the chance for new insights, appear at all stages—including when a scientist and groups of scientists are first researching a problem and when they are delivering their findings. Yet, much of what we have been looking at in the previous chapter has focused exclusively on rhetorical problems that appear primarily at the research stage. These problems inhere at the single word level, or at the level of a cluster of words. As we have seen, the elusive meanings and harmonic implications of metaphoric terms like "signaling" and "fitness landscape" have proven to be a challenge for scientists attempting to represent knowledge. These issues of semantics are rhetorical, of course, because scientists ponder word choice according to the purpose they want that word to serve in a theory, and with an awareness of how an audience will respond to that word.

It is in the stage when words are combined into a journal article or other form of output that problems of rhetorical style beyond the word level become most apparent. The challenge is in finding the best available means of persuasion, to use Aristotelian terminology, while still remaining faithful to the knowledge being presented. These are questions of eloquence, of how a rhetor (in this case, a scientist or technical writer working with a scientist) can offer pleasing and persuasive prose that carries the message without dominating it. While the previous chapter considered primarily problems of rhetoric as a tool of knowledge-construction, this one will deal more with rhetoric as a tool of argument—recognizing, of course, that the two functions are intertwined.

Metaphor is a major rhetorical focus at the Santa Fe Institute. As we have seen, the struggle over which metaphors to use in theory building carries though to the delivery stage when audiences are asked to embrace terms like "epsilon machine" even though no machine is involved. Yet, obviously metaphor is not the only language issue that concerns Institute scientists. Other rhetorical challenges that scientists discussed during the interviews relate to the way they present their findings to other scientists and to the world. These involve challenges of style and eloquence, and those of communicating with scholars in different discourse communities and with the lay public. This is the problem of "incommensurability." It is the topics of style, eloquence, and incommensurability that we will turn our attention to now. These are some key points I hope the reader will take from this chapter:

- Scientists are aware that writing is a balancing act. They must be faithful to the theory they are attempting to explicate in writing, even when that theory is abstract and elusive. But they also must present their insights in a manner that is understandable and engaging for the reader.

- A researcher or technical writer of science must remain faithful to his or her purpose in writing, which is to present an honest intellectual exploration that is focused on the research. In doing so, the writer of science will most likely strike the best balance between substance and style. We could say, using a metaphor, that the scientist or technical writer should not write beyond the limits of the research any more than a driver at night should drive faster than the distance his headlights will allow him to see.

- The rhetorical challenges may change when a scientist moves from the theory-generating phase to the application phase of her writing. At the theory-generating stage, the issues of word meaning dominate, as we have seen. A scientist may employ esoteric terms in a way that would be confusing to anyone not intimate with her thought process. At the application stage, a scientist tends to adopt a more universal lexicon and focus more on global rhetorical concerns, like style, which are above the individual word level.

- Eloquent science writing acknowledges the harmonics of metaphor and manages them in context of the larger theory, so that the overall effect is concordant.

- Scientists have always been wary of any writing style that is overly attentive to personalities and narrative, perhaps finding it too subjective. This instinct is strong at the Santa Fe Institute.

- Scientists, like any writers, must always be aware of their audience. Audience concerns may be more acute than normal at the Institute, where people from different disciplines collaborate on work for journals that are often outside their specific disciplines. Several SFI scientists voice frustrations at trying to strike the right tone when writing for these scientific journals.

- Scientists and technical writers working with scientists can serve multiple audiences, but to be effective they should return to the rhetorical invention stage every time they attempt to rewrite their work. Writers of science taking a realist's view of language and science (maintaining, as we recall from Chapter 4, that scientific theories can accurately represent reality and not merely harness it) might be inclined to believe that the facts or data will shine through regardless of how the text is customized for individual audiences. Holding such a belief, they might attempt to make their work accessible to the lay public not by recreating new images of their audiences, but merely by reconfiguring arrangement and style to convey difficult material and make the facts stand out more clearly. As our brief foray into two texts by Stuart Kauffman will reveal, attempting to serve new audiences primarily by modifying style is a risky strategy.

- Incommensurability is a challenge at an Institute, when scientists are considering a "fact" about the world. Often it takes a lot of time for two researchers to come to understand how each other views that fact, or as Wittgenstein would say, to learn how each other is playing the language game.

- Academics and scientists are wary of colleagues who attempt to cross disciplines, or of programs that would require them to view the world from various perspectives. One reason for the suspicion may be that those scientists are afraid that an interdisciplinary foray would needlessly divert them from their real business of securing grants and winning tenure. Still, the interdisciplinary approach is gaining acceptance among mainstream science funding agencies. This suggests that think tanks such as the Santa Fe Institute may have participated in a "paradigm shift" that is filtering down to "normal" science.

STYLE AND ELOQUENCE IN SFI WRITING

Scientists such as Francis Bacon have been suspicious of words, especially when those words are figures of speech and other seeming embellishments of style. Recall Bacon's exhortation to beware of the "idols of the marketplace," where "vulgar" language and imprecise terminology corrupt understanding (89). The ascendancy of mathematical sciences could be viewed as an attempt to protect theory from the vagaries of discourse. Cowan and other SFI scientists have suggested that some scientific realities can only be expressed mathematically, implying a purity of mathematical language. Other SFI scientists have suggested that words can approach the purity of mathematics if they are chosen carefully—"effective variability," for example, instead of the looser term, "uncertainty."

Still, once a scientist uses words, however carefully, he invites consideration of eloquence and style. Even mathematical equations are subject to questions of style; otherwise there would be no need for various books to teach novice mathematicians how to write proofs. Formulas are kinds of sentences chosen to impart meaning. Style questions stir up ancient tensions over whether eloquent discourse is honest, inspired, and successful discourse or whether it is deceitful discourse. Plato seemed to believe that the truths of the world were so delicate that any expression of those truths in language would damage their truthfulness. For Roman rhetors, the challenge of eloquent writing was the challenge of *decorum*, which entailed the skillful linking of substance (*res*) and style (*verba*). The *Rhetorica Ad Herennium*, attributed by some scholars to Cicero, suggests that eloquence is the use of rhetorical figures to "confer merit and character on language" (quoted in Fahnestock, *Rhetorical Figures* 7).

Defining "eloquence" is not easy, but every literate person can recognize what he or she would call "eloquent." Darwin and other founding thinkers in science developed a style that was literary and fun to read, yet insightful. Current works by Santa Fe Institute scientists and associates like Johnson, Kauffman, Casti, Arrow, Shalizi, and Gell-Mann seem to balance the requirement for scientific rigor with that of eloquence. Still, it seems as if some science writers strive to produce prose that is devoid of pleasure to ensure that it is correct. We will return now to the interviews to see how SFI scientists responded to one of the questions I posed: "Should scientific writing be eloquent, or does eloquence lead to distortions of meaning?" My analysis of these responses is by no means as detailed or imbricated with philosophical considerations as we saw in the previous chapter, but it does offer insights into the problem of style-as-argument in science.

As is the case with metaphor, most SFI scientists in their interviews recognize a need for eloquence in science writing. They also recognize the challenges eloquence poses to the writing process, especially when the desire to be eloquent intrudes on the need for a faithful representation of reality. For Cowan (SF), making nature more understandable is a form of eloquence. A scientist who has "a rich mind and relatively poor command of language" is at a disadvantage in communicating ideas, Cowan says. Yet, he wonders if eloquence is the right word for making ideas accessible, suggesting, like Plato, that eloquence may be going beyond what is required to represent reality. Cowan says his writing is not eloquent. "I try to use words in a very precise way, more precisely than the average person does. . . ." For Cowan, eloquent writing often masks a lack of knowledge—especially in the writing of science journalists. He laments that words "have become plastic" in our society, which suggests that the challenge of representing knowledge is becoming increasingly difficult.

Arrow (SF) also likes writing that is pleasing and flows well, so long as that pleasing style does not come at the expense of accuracy. "I like style," Arrow says. "Some people's writing you enjoy. Some you say, 'Oh my God, I know I am going to learn something, but it's going to be awfully painful.' " Like Cowan, Arrow seems to make a Platonic distinction between what is faithful to reality what is stylistically pleasing. "When the choice comes between . . . clarity and accuracy on one hand and eloquence on the other, you've got to come down on the former," Arrow says. In this sense, style should almost be invisible and should not outshine the scientific reality being represented. "The good writing has to come as a byproduct," Arrow (SF) adds. He defers to advice attributed to Winston Churchill (probably originating with Samuel Johnson) on how to handle eloquent language that rises above the message it is carrying: "After you've written an essay, read it over carefully and if you find any expression that you think is unusually fine, cut it out."

Alexander picks up a theme that Cowan introduced, which is that scientific writing and oral presentations are at their best when done by people who are in command of a subject. "My impression is that the people who know the subject best are clearest and easiest to understand," Alexander (SF) says. In this sense, the writing can be eloquent, but neither affected nor deceitful. Substance and style are blended decorously. Yet, while Cowan is most concerned with overly profuse and inaccurate journalistic science writing, Alexander's concern is more with overly turgid science writing, difficult language "done to impress other people in the audience . . . the three people out there who know the subject extremely well." Alexander suggests that the most troubling science writing is that which is abstract. Hence, she finds the work of another Institute economist, Brian Arthur, to be easily understandable "because it is applied." (Arthur considers how uncertainty and information affect the stock market and other economic arenas.) Alexander's comments focus again on the problems of language as a means of representing reality, by suggesting that ambiguities of meaning resolve themselves when a theory is reshaped to fit an application. This is a kind of pragmatist or instrumentalist assertion that meaning inheres in the tangible results or outcome of an idea.

Moore (SF) also dislikes an overly formalized style of scientific writing, which implies more certainty, objective distance, and profundity than usually is justified. "One annoying thing is that academic writing often obscures the intellectual process," Moore (SF) says. "The journals, I think, encourage you to present your results as if they sprang full blown from your head." These comments are reminiscent of observations that many scientists are Platonists who believe that pure truths exist, and that the goal of science is to uncover those truths without staining them with one's personal beliefs. Indeed, in much SFI writing that reports on abstract theories modeled by computer simulations, it often is difficult to tell much about method or about the process of finding knowledge—who did what. As Latour and Woolgar showed in their ethnographic study of a biological laboratory in the 1970s, the purpose of science is to assay facts, not to tell stories.

Alexander notes that scientists are wary of any writing style that is overly attentive to personalities and narrative. Perhaps it seems indulgent. So a book like that written by Stanford anthropologist Stefan Helmreich that looks at the culture of artificial life sciences is dismissed by some SFI scientists as being subjective. Helmreich follows the current conventions of ethnographic writing by describing his interactions with SFI researchers; this is a technique I have adopted throughout this book. But a writer's intent to inject honesty into the process of gathering knowledge may appear to some scientists as self-centered. Scientists are "uncomfortable" with Helmreich's book, Alexander (SF) says. "He talked about his own biases and his own background, and to some people that may have sounded too personal and not professional. He couldn't, he didn't distance himself from his writing and his work."

Arrow cites the 1992 book, *Complexity,* by M. Mitchell Waldrop as being guilty of some of the same rhetorical sins that Alexander pointed out in Helmreich's work. For Arrow, Waldrop focuses too much on the personalities involved in the founding days of the Santa Fe Institute, and he promises too much from the new complexity science. "Waldrop's book in particular is really a little gushy," Arrow (SF) says. "I think it was an unfortunate book for SFI, actually."

The goal for scientists would seem to be striking a decorous balance between plain science writing and an accessible style. Such eloquence is evident in the writing of SFI biologists Michael Lachmann and fellow researcher Carl T. Bergstrom in their working paper on the cost of signaling in biology, "When Honest Signals Must Be Costly." For example, the two employ figurative devices such as this rhetorical question: "Must there be a cost associated with telling the truth, as well? In Grafen's model, the answer is 'yes'"(2).

Another example from the same paper is the use of *antimetabole,* a figurative device that develops contrast by reversing word order:

In practice, however, many traits involved in advertising will have a utility outside their role as a signal: some may be primarily signals with some functional consequences, while others may

be predominately functional traits that serve secondary signaling
roles . . . (3).

This *antimetabole* successfully manages two words rich with harmonics, "signal" and "function," by showing how they can sound convincing when used together in a theory regardless of which one dominates
the rhetorical relationship and provides what we might call "the primary tone." Lachmann (SF) says that he uses acceptable scientific language, but he tries to make it as pleasing as possible without straying
beyond the results of research. "The language itself, we try to write it
so that it is as clear, beautiful, and understandable as possible,"
Lachmann says. "So it could be the case that we want to say much
more than is said here, but we try to limit ourselves. Scientific writing, he says, must "somehow follow from the results. . . ."

Lachmann implies in his comments that a proper balance
between style and substance will follow if a researcher remains faithful
to her purpose in writing up the findings, which should be to present
an honest intellectual exploration focused on the research. Plato's fear
of rhetoric, we may recall, followed from his assumption that the
rhetorician sets out to be persuasive. Certainly if the purpose of writing is persuasion, it is easy to see how a writer could stray from sin-
cerely presenting the results of his studies. But Aristotle later would
challenge the Platonic inference that eloquent writing is the product
of intent to persuade. He wrote in *The Art of Rhetoric*, "It is also clear
that its [rhetoric's] function is not persuasion" (69, italics original).
Aristotle's rhetoric is philosophical, even dialectical, in that it
approaches a problem by assembling relevant proofs until the problem
is brought into focus. The process of bringing a scientific problem
into focus, of course, will be persuasive, leading to probable conclusions, but the intent should not be persuasion before truth. Arrow
(SF) arrives at this conclusion about persuasive writing when he says,
"The good writing has to come as a byproduct, I think."

Good science writing, then, would function as a philosophical
process aimed at finding the best representation of reality as a scientist
sees it. That writing would not hide the paradoxes inherent in many
scientific terms, such as "signaling" or "pattern," but would engage
those paradoxes honestly. Shalizi and Crutchfield in writing about
computational physics appear to have engaged the paradoxes in their
research in a way that is honest, and therefore, convincing. At the
close of their working paper on patterns, they address limitations of

the research, admitting that many things remain "unclear" and that some language is "vague" (18–19).

Perhaps eloquence in science writing is desirable if it is clear that the writer's purpose is not overt persuasion, but a search for the most faithful representation of reality possible given the constraints of research and language. Casti, who has written many popular books related to complexity sciences, reveals how a sense of intellectual purpose motivates his approach to writing. "Almost always when I write it's because I'm trying to understand something myself that I don't understand, or feel that I understand well enough," Casti (SF) says. "And I'm of the belief, for me anyway, that one of the best ways I can gain understanding is try and explain it to somebody else who may even understand less than I do." Casti says he is not proselytizing in his writing or trying to "start a new religion," but just trying to work through how he understands difficult concepts such as "chaos." We have seen that rhetorical choices at the individual word level are theory constituting. Casti clearly is making the case that rhetoric at the sentence level and beyond also constitutes theory. He is writing to learn.

Clearly one of the most demanding rhetorical challenges scientists face is when explaining their work to the lay public. They may be forced to simplify mathematical concepts such as chaos in an effort to explain the significance of these concepts to the world at large. As Dulle (SF) notes, scientists who develop impressive models often have a difficult time answering the question, "And so what?"

Shalizi seems comfortable presenting abstract and difficult topics in an accessible style. His paper on patterns in physics, co-written with Crutchfield, blends philosophical and literary references with the algebra of information theory. Shalizi says he came from "an intensely intellectual family" and has always enjoyed scientific and philosophical writing. Like others, he asserts that it is possible to strike a balance between pleasing prose and scientific fidelity to the observed world. "If it's only eloquent then obviously one is justified in becoming suspicious that, you know, somebody's trying to put something over on you" Shalizi (SF) says. "But there is no reason why you can't be eloquent, why you can't write something which will be just pleasant to read or hear, which is also factually and logically sound."

Fontana (SF) describes the balance between substance—observed reality—and style as "disciplined" eloquence. Such a balance is essential, he says. "There is no point in having theories that you can't communicate," Fontana says. You can't force someone to read

100 pages of "detailed gritty stuff" just to find out if they are interested in the topic. Therefore, it is necessary to compromise some of the details of one's theories in exchange for readability. As Fontana describes it, "You have to sort of cartoon your own theories to some extent, and to capture in an accessible way their essence." Still, Fontana says that he has difficulties communicating what he is trying to get across. "Writing to me is an extremely tedious process," he adds. These writing challenges are not unique at the Institute, as we have seen from Johnson's accounts of Gell-Mann's struggles, or among scientists in general. Johnson (SF) recalls talking to editors of various scientific journals who have told him of the "agony" they go through trying to turn copy from scientists into readable articles.

For Shalizi (SF), one way to avoid suspicious eloquence is to avoid terms that are loosely defined, or "flaky." This again relates to the problems of metaphoric language that is perhaps rich in implications, but impossible to pin down to a core meaning. Recall Morowitz referred to such terms as "meta-metaphor," which he saw as dangerously vague. One such phrase might be "edge of chaos," which we have seen used to describe the condition between entropy and order where life and other interesting phenomena occur. Another phrase Shalizi would avoid as being imprecise is "complex adaptive systems," he says. "There is no good way of distinguishing what one is." He recalls once starting to write a tongue-in-cheek paper with a colleague arguing that neutrino reactions in a supernova could be examples of a complex adaptive system. "And, you know, this is complete nonsense of course, because there is really nothing adaptive in there, and so on, but the way it's used, we could have done it," Shalizi says. His comments recall the famous Sokal hoax in the mid 1990s, when a physicist was able to publish what turned out to be a spoof article in a literary journal proclaiming that gravity was entirely a linguistic construct.

SFI scientists, like others in their profession, must develop a sense of when certain terms or stylistic devices may be appropriate and when they are not. Academic journals typically require staid language that is more dry than evocative. Scientists would be prudent in avoiding a phrase like "edge of chaos" when writing for the *Journal of Theoretical Biology* or *Communications of Mathematical Physics.* Economist Shareen Joshi (SF), for example, says she feels free to use concepts like "edge of chaos" when envisioning the economy in her mind, but she would never use such phrases in a journal article.

"They're useful to me in terms of how I think about things," Joshi says. "But other than that, I'm cautious because there's been a whole lot of skepticism, a whole lot of criticism in the past few years." The main problem with such phrases, Joshi, notes, "is they don't have concrete definitions." Awareness of audience need for clear definitions also determines how Casti balances style and substance. For formal science articles, Casti (SF) says he will try to be "a little more precise," with less "fuzz around the edges."

Several SFI scientists voice frustrations at trying to strike the right tone when writing for an academic journal. "A lot of people wish the journals would go away," Moore (SF) says. Moore says he is pleased to see scientists writing up their results on various Internet Web sites instead of traditional journals. His comments suggest that young scientists find the formal style of journals and their careful review process to be intellectually stifling. The Internet is encouraging more informal writing that is more "playful," Moore says. "There is just no excuse for a journal to take two years to publish results." Yet, Moore qualifies his comments, by stating that journals serve a purpose in filtering the tremendous volume of academic writing that scientists produce. Perhaps at their best, the journals serve to ensure a prudent balance between substance and style in scientific writing. Of course, journals maintain the standard that rising academic scientists must meet if they are to be promoted into the society of their peers. Knowledge that appears in journals is, therefore, social as well as cognitive.

Roger Turner, a student of intellectual history and the history of science who has spent some time at the Institute, argues that the pressure to publish may keep new ideas flowing in science. No one school of thought can dominate, Turner (SF) says, because scientists who are always trying to get published are forced to take fresh perspectives. "At least in modern science there is such a pressure to publish, and there are so many different perspectives, where one school of thought's perspective is often consciously defined in opposition to that of another school of thought," he says. "And so it seems like just because something has been written doesn't necessarily make it as authoritative, at least within science."

From these brief comments on style and eloquence offered by Institute members, it seems obvious that scientists are aware of the tensions between style and substance. In the interviews, I did not have to explain much rhetorical theory to prompt responses to these questions.

Undoubtedly these scientists consider rhetorical challenges frequently as they write up their work. All of those who spoke about eloquence appear to believe it is important in science writing, and that possibly science could use a little more eloquence. But all maintain that eloquence that runs too far ahead of the data is dangerous. This challenge of choosing an appropriate rhetorical style seems most acute when scientists write for an audience of peers in an academic journal, or as we will now see, when trying to rewrite a work for a new audience.

A MATTER OF INVENTION: WRITING FOR SCIENTISTS AND THE GENERAL PUBLIC

Stuart Kauffman has written two books explaining his detailed research and ruminations into the tendency for pre-biotic chemical agents to self organize into enzymes, genes, and other components of life. *The Origins of Order* clearly is scientific writing that Kauffman has directed at an audience of peers. *At Home in the Universe*, found on the shelves of popular bookstores, is directed toward a more general audience. Both books expound upon Kauffman's central argument developed over the course of his life—an argument we have seen in various forms throughout much Santa Fe Institute research. Kauffman maintains that order and increasing complexity in life are not merely the products of an evolutionary natural selection process that mechanically propels species to greater levels of fitness. Order also emerges, Kauffman argues, as the components of life interact with each other and achieve synergies that would not be possible in confinement. But this "order for free," as Kauffman calls it, comes at a hidden price; the interaction of the agents of life change the environment, the playing field or "fitness landscape," on which those agents interact. So rules for success change with evolution and interaction.

The Origins of Order, published in 1993, spends about 650 pages exploring in detail the tensions between emerging order that arises among agents on the landscape, and natural selection, which determines whether the new varieties of ordered life will succeed. Chapter and section titles include, "The Structure of Adaptive Landscapes Underlying Protein Evolution," "Morphology, Maps, and the Spatial Ordering of Integrated Tissues," and "Random Grammars: Models of Functional Integration and Transformation." While this book has its moments of rhetorical flourish, even Kauffman acknowledges in a follow-up interview that parts are tough going and unwieldy. "It's a massive tome of a book," Kauffman (SF) says. "A friend of mine calls it 'the beast.'"

Kauffman in 1995 offered a more popular and accessible account of the same material, employing the close editing assistance of *New York Times* Science Writer George Johnson, whom we also met in the last chapter. Spanning half the length, even the title, *At Home in the Universe*, sounds less profound and less self-important, albeit folksier and more spiritual, than the original. Kauffman pulls out all the stops and writes with rich metaphors, narrative digressions, and rhetorical hyperbole. Chapter titles are more appealing; they invite the reader to join Kauffman in a rhetorical *paenismus*, or exclamation of joy—in this case, almost the spiritual joy of being alive. So instead of the formal scientific headings of *Origins* we have "We the Expected," "Noah's Vessel," "The Promised Land," and "High Country Adventures."

Origins is full of mathematical formulas, in-text citations, and careful clarification of terminology. The popular account appears more narrative, hyperbolic, poetic, and full of epideictic wonder. The first chapters of *Origins* and *At Home* both contain histories of evolutionary theory, although the technical book leads with a formal declaration of intent for the chapter and proceeds straight to the literature review. The latter leads with a narrative of Kauffman and other scientists contemplating life's origins from an overlook at the Santa Fe Institute:

> Some months ago, I found myself at lunch with Gunter Mahler, a theoretical physicist from Munich visiting the Santa Fe Institute, where a group of colleagues and I are engaged in a search for the laws of complexity that would explain the strange patterns that spring up around us. (*At Home* 3)

A close look at key arguments in the two books reveals interesting differences, which point to a writer trying to revise a work for a new audience. Kauffman's central metaphor in both books—indeed the central metaphor for Kauffman's scientific worldview—is the concept of "fitness landscape." This we have seen describes an environment of living agents that evolve to new levels of fitness as they interact over time. So we imagine the peaks and valleys of this landscape changing as the agents become more or less successful at survival and reproduction. In *Origins*, Kauffman introduces the metaphor formally by acknowledging its source, a biologist working in the 1930s, but without providing a succinct definition of the metaphor. Instead, Kauffman moves immediately into a theoretical discussion of the relationship between fitness landscapes and co-evolution (*Origins* 33). Such a move is typical of formal academic writing, where paying homage to a predecessor is often all that is necessary to introduce a concept to an audience of peers who presumably would be familiar with it. Yet, the predecessor is referenced in *Origins* only by his first initial and last name, "S. Wright." Formal academic writing, as we know, attempts to downplay the cult of personalities; hence a real biologist in the 1930s is acknowledged, but almost as if he were a fact himself associated with a theory rather than a person with a full name and a life story.

In *At Home*, however, Kauffman defines "fitness landscape" and then offers an example of how such a landscape would be found in the environment of protein molecules. Another interesting distinction between the treatments of the fitness landscape metaphor is found in the two terms Kauffman uses to introduce it: In *Origins*, a fitness landscape is referred to first as "an idea," (10), while in *At Home* fitness landscapes are referred to as "images" (26). Seemingly, the more academic treatment elides the *sensation* of a metaphor and focuses on its epistemological function, while the popular treatment foregrounds the metaphor as an image that can be seen and experienced, not merely known.

In *At Home*, Kauffman introduces the concept of a fitness landscape without attributing it to Wright. But later in *At Home* we meet various founding geneticists, including Wright—now introduced by his full name, Sewall Wright (*At Home* 162). Personalities have a more prominent role in popular science writing, where writers of science face less pressure to reduce all information, even information about people, to objective data.

These obvious differences in the two books prompted me to seek the follow-up interview with Kauffman (SF). The intent was to combine a reader's analysis of two texts with the author's insights into the rhetorical choices he made when writing each. From this exploration, we can see that a scientist's consideration of audience continually shapes the way theories appear as they form out of the writing process. The argument that emerges from this analysis and author interview is that a scientist must reinvent a theory when revising any text that attempts to deliver it to a new audience. Even when the author is not entirely successful at reinventing a theory, a new consideration of audience will invariably cause that theory to change not only in style, *verba*, but also in substance, *res*.

Kauffman explained that the science in both books is almost identical, but that the two books are for "two vastly different audiences." Kauffman's comments reveal that he had specific audiences and purposes in mind when writing both books. *Origins*, he says, was directed at an audience of academics with their Ph.Ds who have tenure at a good university. "There's no doubt when you read it that you are reading a scientist who is writing to scientists . . .," Kauffman says. His purpose in writing *Origins* was equally focused, "to persuade a massive professional audience of the necessity to rethink evolutionary theory."

While Kauffman had a group of peers in mind for the first book, he had an even tighter audience focus for the second. "I had a very specific reader in mind and it's Al Gore," Kauffman says. "I mean quite literally, I wrote the book so that Al Gore would understand it." Gore had visited the Santa Fe Institute in the 1990s during his tenure as U.S. Vice President. As Kauffman recalls, Al Gore is "real smart . . ., but has no specific schooling in the sciences." Hence, Kauffman set about explain the science in "layman's terms." Yet, he adds, this audience of lay people accompanying Al Gore in the author's mind could be clearly delineated. It would comprise people in "the top 5 percent of the national population," people with a measurable intelligence quotient of 130 or above. The purpose of *At Home*, Kauffman continues, was "to sort of stun, amaze, and delight" this gifted group of readers.

Kauffman's comments strongly support research in the rhetoric of science, which makes a convincing case that scientific language and the facts it conveys shift significantly according to the author's perception of who will be reading. Audience is so important to rhetorician Phillipa Jane Benson's analysis of the scientific writing that she uses different terminology for that writing according to the intended audience. For Benson, "scientific writing" is directed from one scientist to other scientists for the purpose of persuading them that the research and interpretations are sound (211). "Science writing" is directed at popular audiences for the purpose of "informing and entertaining" (212). Here we see a restatement of Cicero's categories of the aims of rhetoric—to persuade, to teach, or to delight. So we could safely conclude that *Origins* is an example of scientific writing, while *At Home* is an example of science writing.

In each case, audience clearly determines the purpose. Recall Fahnestock's research, which reveals how sciences that are especially relevant to human life, such as biology, are represented by different language depending upon audience. Biologists writing to a small group of fellow scientists couch their claims in modifiers, offering new findings in a forensic style. That is, new discoveries are presented cautiously for the judgment of scientists. They are draped in hedging terms like "it seems likely" and "our data are consistent" ("Accommodating Science" 28). The same discoveries when popularized appear in a more epideictic style, a laudatory praise of the wonder, surprise, and potential embodied in the discovery.

Kauffman recognizes that *At Home* is full of poetic expressions of awe and reverence for the universe. "I wanted it to convey the sense of us as members of a creative universe," Kauffman (SF) explains "And in that sense, it's spiritual and religious. There's no god in it; there's no supernatural force in it. There's membership in a creative universe—that is the spirituality for me." Kauffman recalls that in his youth he enjoyed writing poetry; hence in *At Home* he allowed himself "the luxury of being poetic." Perhaps Kauffman went too far at times. At the end of the writing process he and Johnson cut out much of the poetry that he saw as excessive, "what we came to call 'the Kauffman purple.'"

In *Origins*, Kauffman offers careful biological description where in *At Home* he might offer poetic imagery. For example, in describing similarities in structure across different species, Kauffman writes in the academic book, " . . . the homology between the reptile jaw and the mammalian inner ear" is evidence of "invariant relations" (*Origins* 5). By contrast, *At Home* offers more accessible examples of structural similarities and treats them with more epideictic awe, significance, and rhetorical flourish. Notice the alliteration and consonance here describing the same science: "Surely the pectoral fin of a fish, the bones of a petrel's wings, and the flashing legs of a horse were expressions of the same deep principle" (*At Home* 7).

Both books are filled with metaphors, which is not surprising since we know that Kauffman and other complexity theory scientists are intensely metaphoric. But the differences in how metaphors are used reveal a writer who is consciously attempting to modify his rhetoric in order to address different audiences. In the more academic version, metaphor is used almost apologetically. For example, when he explores the structure of landscapes in *Origins*, Kauffman writes of "local hill climbing via fitter mutants. . . ." (33). He then seems to apologize for the limitations of the image: "Despite this *transparent* metaphor, such a process involves complex, combinatorial combination" (33, italics mine). This is consistent with what we have seen where scientists use metaphors for their powerful effect, while acting ashamed of this dependency.

In other cases in the academic book, Kauffman uses metaphors as formal scientific terms, without calling attention to their presence. The relationship between the number of agents (N) and how often they interact (K) allows for a model "of an ordered family of tunably correlated landscapes," Kauffman writes in *Origins* (54). In *At Home*, Kauffman makes the metaphor explicit by opening it up to reveal the simile: "Altering K is like twisting a control knob" (173). As Johnson notes, the version in *Origins* "might be an example of something where to Stuart it wouldn't be clear that you needed to explain what tunable is." Once Johnson suggested that the metaphor needed to be unpacked, Kauffman willingly made the change for the second book. Yet, in making more explicit the concept of "tuning," Kauffman also reveals some theoretically troubling aspects of the term and its harmonics. Tuning a control knob requires some higher entity, although presumably agents on a fitness landscape would be autonomous members of a self-organizing system. The images of a natural landscape and machine control knob clash. Perhaps this discordance would not be apparent to the scientific audience encountering the control knob metaphor buried in the formal text, while the lay audience encountering the same metaphor more overtly would lack the precision necessary to hear its overtones.

Anecdotes from Johnson and a close reading of the text of *At Home* reveal that Kauffman may have had difficulties keeping his audience in focus when moving back and forth between this and the earlier academic work. This focusing difficulty is understandable given the panoramic scope of Kauffman's science. Johnson reports that Kauffman experienced severe writer's block in producing *At Home*. He recalls telling Kauffman to write what came to him and then send the draft to Johnson, who would go through the book line-by-line suggesting improvements. The original version of *At Home*, Johnson recalls, was "so long . . . incredibly dense." Is it possible that

Kauffman was still writing to the same audience of peers he envisioned for his first book?

Consider the following examples of superficial changes in style that did little to embrace a new audience. In the technical book we find some of the terms of condescension toward non-experts that often fill formal scientific texts. Kauffman concludes a difficult passage dealing with genes, alleles, and fitness landscapes by blithely proclaiming that, "*It follows trivially* that there are no optima other than the single global optimum." (*Origins* 45, italics mine). This is softened in the popular version to, "*So* there are no other peaks on the landscape, for any other genotype can climb to the global optimum" (*At Home* 174, italics mine). Introducing an earlier passage dealing with the same topic, Kauffman writes, "*It is easy to see* why increasing K increases conflicting constraints" (*At Home* 173, italics mine).

One could point out, as Johnson might have done to Kauffman, that the softer language and tone in the popular account indeed show sensitivity to the lay audience. Yet this material is no more "easy to see" than it is "trivial." Kauffman has attempted to touch up the rhetoric, but without clearly envisioning a new audience. If he had envisioned that audience, he would have made the argument in the ensuing passage easy to see instead of merely proclaiming it to be so.

In another passage in *Origins*, Kauffman explains an accompanying diagram of a fitness landscape using specific terms from biology: "Figure 2.2c shows such a fitness landscape for eight possible genotypes available with three genes, each having *two alternative alleles* (*Origins* 43, italics mine). In *At Home* the explanation for the same figure, numbered differently in the new book, removes some of the biological terms, but adds new symbols from combinatorial mathematics: "Figure 8.5 shows an example with N=3 and K=2—that is, the genome has three genes, *each of which is affected by two others*" (*At Home* 172, italics mine). The change makes part of the new sentence more accessible, but makes another part more difficult.

Kauffman's popular book often moves from philosophical speculation about the nature of life to short refrains, such as, "We the unexpected; we the very lucky" (*At Home* 149). This kind of litany delivers his argument in an oral style that is easily accessible and easily remembered. Yet, when he moves so quickly from these aphorisms to specific scientific detail it becomes apparent he is not keeping one audience in focus. "It is one thing for me to tell those of you who are not biologists that self-organization plays a role in the history of life,"

Kauffman writes (*At Home* 150). He adds that this seems obvious, but then after having just addressed non-experts, Kauffman supports his claim with an example that would be accessible primarily to biologists: "After all, lipids in water do form hollow bilipid membrane spheres, such as cell membranes, without the benefit of natural selection" (150).

Kauffman defends this choice of wording in his follow-up interview, arguing that Al Gore would "understand perfectly well that if you took some lipids and shook them up you'd get . . . a hollow ball that was like a lipid membrane." Yet, it seems obvious that Kauffman is confusing an audience of Gore and other intelligent lay people with one of scientists—the same group of peers he was writing for in *Origins*. To wit: I spent several hours in 1997 with a group of college-educated businesswomen at an El Paso, Texas book club discussing *At Home*, which was their selection for the month. Few of these readers got much more than a fuzzy sense of Kauffman's argument.

It seems that scientists and technical writers of science can serve multiple audiences, but to be effective they should return to the invention stage every time they attempt to rewrite their work. Admittedly, this invention process would be difficult to follow. Doing so would require writers of science to repeat much of the writing process whenever they seek to present work to a new audience. Yet, it may not be enough to simply change the order of material in a text or the style in which it is written. Clarifying scientific metaphors may not be sufficient to render them useful to a new audience if they were not conceived with that audience in mind.

I should note here that this call to reinvention does not deny that writers of science can draw on the same text and illustrations to serve multiple audiences. This concept, known as "single sourcing" in the technical communication field, assumes that root documents can be incorporated into manuals and other texts that are intended for audiences possessing different skills and needs. Reinvention can involve using existing text in new settings as long as the writer can see the appropriateness of such text from the new audience's perspective. Writing should be an organic process whereby words, phrases, and even whole blocks of text are called into play by an author who is in a deep implied dialogue with her audience.

To explore the claim that writers of science must reinvent in order to rewrite, we must digress briefly back into rhetorical theory as it relates to rhetorical invention. This is the first of the three steps of

Classical rhetoric that are applicable to scientific writing. Invention, which determines the content of a text, often is reduced in writing classes to a series of planning exercises, referred to as "brainstorming" or "pre-writing." Strategies of invention in technical writing classes almost always begin with two questions: "What is the purpose of the writing? And "Who is the audience?" These questions can be collapsed into one, which is, "Why does an audience need to know this?"

In the second and third canons of rhetoric, the real or imagined audience leads the writer to arrange the information and present it in a style that will serve that audience expediently. Different types of scientific and technical writing have different aims, of course—everything from helping computer users to adopt new software, to presenting a new argument about the nature of the universe. Regardless of the individual writer's aim, the most successful technical and scientific writing occurs when the writer can envision a real audience throughout all the rhetorical stages. This process of envisioning audience begins in the invention stage, for this is when the writer chooses the components of an argument that will be organized and adorned in a meaningful and useful way.

Rhetorical invention for Classical theorists was a process by which a rhetor found the content for an argument. Cicero defined invention as "the discovery of valid or seemingly valid arguments to render one's cause plausible" (*De Inventione* I., VII). Narrowly defined, rhetorical invention is simply the retrieval of ideas already extant; hence, it would not entail bringing into being something new. A broader definition, one favored by those taking a social-epistemic viewpoint, is that rhetorical invention is a process of inquiry, similar to that conducted in a scientific laboratory. So the process of finding arguments for discourse or for a text assists in the development of new scientific knowledge. The 1971 Report of the Speech Communication Association's Committee on the Nature of Rhetorical Invention makes such a claim: "Invention (used now as the generic term) becomes in this context a productive human thrust into the unknown" (quoted in LeFevre 3). Therefore, the writer is thrusting into that unknown on behalf of an audience whose members cannot be present to do so for themselves.

Invention as a process of selecting the most effective proofs has never been a solitary process. The ancient Greeks certainly assumed the presence of the "other" in their rhetorical invention. Plato searched for truth by setting up a dialectic process by which real or imagined characters sparred intellectually over some question. Aristotle also employed the dialectic method of arguing from opposites. Invention for Aristotle became a process of choosing among available topics and employing deductive and inductive reasoning based, in part, upon the psychological profile of the audience. Audience was essential, for it often filled in the missing warrant to Aristotle's proofs. Audience also helped to define the moment when an argument becomes ripe for delivery; the Greeks knew this consideration of timing as *kairos*.

Writers, including scientists, are able to determine what to write because they are constantly consulting with others in their field, determining what gaps in understanding exist, what new research is needed. This consultation occurs informally as scientists go about their daily research activities, but it becomes more formal when the scientist attempts to write about findings for a research journal. At this point, the editors function as advocates for the audience by calling upon the scientist writer to make new research relevant to the existing body of knowledge. Greg Myers considers the tensions that emerge with editors when biologists attempt to publish: "I argue that the writing process is social from the beginning, and that there is a tension inherent in the publication of any scientific article that makes negotiation between the writer and the potential audience essential" (188).

Keeping a potential audience focused in the mind's eye, then, is not only a noble rhetorical goal, but also one that is essential for a successful publishing career. Of course, it is much easier said than done. Scientists-as-writers do not develop their ideas fully formed in a single rhetorical setting; rather they incubate and refine these ideas over many years among many different colleagues and acquaintances. These are all potential audiences for their text, and they all have different needs. Scientists in the midst of a research program must write different accounts of their work for fellow scientists in their area of specialty, for scientists with different specialties, for granting agencies, for academic generalists, and for the lay public. By the time a scientist sits down to write a book that summarizes the work of a decade or more, that writer no doubt has related bits and pieces of the research to many different audiences. Kauffman spent many years with many real and imagined audiences developing the ideas that would go into *At Home*.

Rhetors have always had to serve audiences of different backgrounds and needs. As Perelman and Olbrechts-Tyteca show in their text on modern rhetorical theory, it is usually the case "that an orator must persuade a composite audience, embracing people differing in character, loyalties, and functions" (21). The Belgian rhetoricians point out the drawbacks of trying to serve various audiences at the same time, which leads the authors to claim that the most effective rhetor finds that which is common in all audiences, in essence, by creating a "universal audience" (31). Members of such an audience ideally are unified by attention to "self-evident" reasons of "timeless validity" that transcend time and place (32). Of course, this would be an ideal audience, but it is one that few scientific and technical rhetors, who deal in specifics, would be likely to encounter.

Instead, the scientist typically encounters multiple levels of expertise in people of different backgrounds. Not all will be persuaded by the same aspects of reality because not all will be able to comprehend the same aspects. Different audiences will hear key words like "tunably" differently; some audiences may be more sensitive to the harmonics that these metaphors carry. Mathematical proofs about the nature of space-time will mean little to someone not fluent in higher mathematical symbolism, but conversely, the same arguments presented metaphorically will seem soft to the committed mathematician. Various scholars of rhetoric—notably Toulmin—have referred to this problem as one of "incommensurability" among different audiences. Still, envisioning a specific audience is not easy, as rhetorician Walter Ong made clear in "The Writer's Audience is Always a Fiction." Ong holds that writers create an image of an audience in their minds playing some role. Readers, in turn, slide into that role if they are to get the writer's message.

A writer of science might be likely to rebut Ong by arguing that the facts, not the author, will call forth the right audience—presumably those readers capable and ready to receive those facts. This implicit realist belief that facts can shine through the text to reach multiple audiences often underlies science writing. Scientists holding such a belief might attempt to make their work accessible to the lay public not by recreating entirely new images of their audiences, but merely by reconfiguring arrangement and style to convey difficult material and make the facts stand out more clearly. This, however, is a risky shortcut that cannot substitute for reinvention.

Yet, it is a shortcut that sometimes leads complexity science writers like Kauffman to go astray. Because scientists at the Santa Fe Institute are working on theories that have such a powerful pull on society, these writers are often lured to present their research to an eager popular audience when they may not have been in an imaginary or real dialogue with that audience during the theory invention stage. Johnson (SF) recalls a period in the mid 1990s when book agents were soliciting everyone who worked at the Institute. "There was this period when this agent . . . swept through the Institute and signed up all of these complexity books," Johnson says. "And one was Stuart's book, *At Home in the Universe.*" Before bringing Johnson into the writing process, Kauffman struggled to invent a new account of his book for this audience, Johnson recalls. "He hated the whole result, hated the process," Johnson says.

The two books ponder the same facts of science, yet the theories that emerge clearly are shaped by the audience considerations. Differences in these theories can be found by looking at the book titles and at subtle differences in the closing paragraphs of each. *Origins of Order* develops the theory that agents interact, causing changes in each agent and an ordered relationship among them, which leads to systems of higher complexity. In *Origins*, the source of such evolution is "a complex combinatorial optimization process in each of the coevolving species in a linked ecosystem . . ." (644). *At Home in the Universe* concludes, however, "organisms are not contraptions piled on contraptions all the way down, but expressions of a deeper order inherent in all life" (304). The key phrase here is "deeper order," which Kauffman in the same paragraph rephrases as "deep natural principles." The theory of this deeper order in this more popular book is captured in the oft-repeated phrase, "we the expected." Kauffman is giving his popular audience "deep natural principles" instead of "a complex combinatorial optimization process"; he is giving that audience a theory of origins that is spiritual rather than blindly algorithmic. The two versions of Kauffman's concluding argument strike the ear differently. Yet, the theory in the popular book is elusive and difficult to hear in the places we have seen where the author does not appear to have reinvented it fully for the new audience.

While *At Home* has been highly praised, several reviewers have pointed to its inconsistent style, which oscillates between sheer poetry and dense science. Johnson notes that Kauffman's writing style in the book has been referred to in one otherwise favorable review as, "Beowulf on bad acid." Kauffman's sweeping grasp of the origins of life and his enthusiasm for complexity theory are staggering mental states to witness as a reader; Kauffman's work in *At Home* pays off the diligence needed to access it. Yet, his work serves a smaller audience than it might have done with more attention to the rhetorical invention process. Perhaps Kauffman's audience for these ideas has always been primarily himself—a polymath who is part poet, part scientist, part philosopher.

INCOMMENSURABILITY: TRYING TO CROSS DISCIPLINES

Trying to bridge the gap between an audience of experts and one of lay people is an enormous challenge for scientists and technical writers working in a field that has broad appeal, as Kauffman's writing saga reveals. An equally difficult task for a scientist is to communicate with other scientists outside her field. The founders who set up the Santa Fe Institute in the mid 1980s did so with the assumption that scientists from various disciplines would benefit from talking among each other and sharing research. They would use reasoned argumentation to adduce relationships among the realities observed in individual fields. Yet, much of the current debate in argument theory asks whether it is possible to use argument skills effectively across disciplines. Toulmin in 1958 first broached this question of field dependency by suggesting that people operate from within logically incommensurate discourse communities. These communities each viewed reality according to what Kuhn referred to as "paradigms," or what rhetorician Kenneth Burke referred to as "terministic screens." Implicitly, Toulmin, Kuhn, and Burke seemed to suggest that people from different disciplines would not see the same patterns in science any more than people wearing different colored sun glasses would see the same tint of reality. Considering the terministic screen metaphor in our auditory analogy, we realize that people from different scientific disciplines will hear words differently as notes appearing in different keys.

Rhetoricians thus argue that paradigms are "incommensurable," which would mean that a biologist would not be able to talk meaningfully to an economist. We have already seen examples of scientific arguments breaking down over the meaning of a single word. Similarly, philosophers like Wittgenstein see language as a game, where one must know the specific rules to have meaningful discussions. Yet, the SFI is premised on the idea that disciplines are commensurable, and that some kind of meta-language can allow people of different communities to argue effectively. Hence, I asked scientists in the interviews how they avoid the problem of incommensurability.

Lachmann (SF) recognizes the challenge of working with another scientist, noting that "it takes a very long time till you manage to understand the other person." This suggests that incommensurability is mitigated only over time. "I mean what I'm saying, but what you hear is something else." He laments that sometimes scientists may think they understand each other, without doing so. "And just talking with people here at the Institute, I talk with them as if they understand me, but I'm sure they often don't," Lachmann says. His response centers on the problem of communication between two people when they are considering a concept that is separate from both of them, a "fact" about the world. If one person speaks of an observation, call it X, there are various stages along the way when the representation of X can change. First, X is represented subjectively by the observer as X_1, and then communicated with approximating language as X_2 to another person, who reconstructs the concept as X_3 in her mind. For those who hold that reality is constructed in these social interactions, perhaps the change of X to X_3 would not be disconcerting; each version of X is as real as the next. But for scientists operating in the positivist tradition, the various guises of reality would be deviant—unless, by chance, each had the exact same predictive power.

Again, the challenge of representing knowledge among people from different backgrounds is in the associations, or harmonics, that words invoke. Of course, sometimes these associations are useful for producing insights. Susan Ballati (SF), who is involved with marketing and development at the Institute, says that a chief executive officer of a business who listens to a talk about ants following pheromones, for example, might conjure up images of the network of distributors who serve his business. Fontana (SF) reminds us that the science of particle physics emerged in the early twentieth century because of an incommensurability problem in which physicists operating out of different research paradigms saw particles either as traces on a film or as a click in the detector. This incommensurability at first brought misunderstanding, but eventually it made possible novel insights into the ephemeral nature of these particles. This "led to the golden age of particle physics," Fontana says.

Fontana's example, drawn from published accounts of the history of particle physics, reveals how one's expectation, or terministic screen, affects the reality one sees. "I think language does shape the way you think," he says. In presenting the metaphor of molecules as mathematical functions to other scientists, Fontana is asking them to shift screens, to avert their attention from one manifestation of reality in order to see another. The physical sciences are mostly concerned with changes in quantities, "observables that have magnitudes" (momentum and energy in physics, for example), Fontana (SF) says. "But there are aspects of the real world where it's not just the quantities that change, but it's the things themselves that change." Things are not only quantities, he says, noting, "molecules interact to produce or change each other."

Although the metaphor that Fontana and Buss developed for molecules as functions generally has been accepted in chemistry and across disciplines, other SFI scientists report that efforts to cross disciplines are not always immediately fruitful. Crossing disciplines is especially difficult when scientists are trying to publish results. If you are a physicist like Moore, for example, who is trying to publish results of research that uses terms from logic to explain the behavior of magnetized atoms, where do you send your article? "I had two papers rejected for a computer science conference," Moore (SF) says. The reviewers wanted his papers to be more integrated with existing work in the field, and to reference established names.

As an economist who had an eclectic undergraduate background at Reed College in Oregon, Shareen Joshi is experienced in crossing disciplines. In some ways, she typifies the interdisciplinary ethos of the Institute. "I've been extremely interested in physics and biology and I read a lot," Joshi (SF) says. "I collaborate with a physicist. I've never really worked with an economist. At Reed my advisor was a philosopher." Her experience has been largely positive, although she recalls when she first got interested in studying issues of complexity in financial markets, her professors at Reed at first were wary. "They saw it completely outside the traditional neo-classical model," Joshi says. "But Reed is a very unusual place. It's an intellectual utopia. And so, even though it was outside the normal and they thought I was probably not making the best use of my abilities, they gave me the freedom to pursue this, and it worked out."

Joshi discovered that although neo-classical economics is based on Newtonian physics, new terms from physics are not automatically welcome in her field. "I used the term 'apparent non-stationarity' as used in physics and an economist told me he didn't know what it meant," Joshi (SF) says. Instead of coining new terms to explain a new awareness in economics about uncertainty and indeterminacy, Joshi has modified the standard term "equilibrium." She uses "disequilibrium" to describe the condition when markets do not clear, that is, settle to a price level that reconciles supply with demand. For Joshi, a term like "disequilibrium" is not a radical departure from conventional lexicon. Similarly, economists who once saw consumers as behaving according to "rational expectations" now qualify their assumption to allow for irrationality—not by abandoning the conventional term "rationality," but by modifying it with the adjective "bounded." Use of the adjective recognizes the limitations of a term as a representation of human behavior, without sacrificing all of the theory imbedded in the original term.

Although economists may be on the lookout for new metaphors to deal with the contingent and uncertain world they study, Joshi's experiences reveal it is exceedingly difficult to throw out established terms and replace them with new ones. Language changes gradually, but it resists decreed changes. Arrow (SF) notes that biological metaphors did not win quick converts to an alternative view of the economy despite the excitement generated by Darwin's theories. "There was a century of this (biology) analogy and really . . . it accomplished nothing," Arrow says. The metaphors were accepted eventually in the 1930s after they could be tested by formal models and mathematical systems. Arrow says that he believes an evolutionary model of the economy will be useful eventually, but "I can't say it's delivered anything, yet."

If borrowing metaphors across disciplines raises eyebrows at a college described earlier by Joshi as "an intellectual utopia," it is not surprising that mainstream science sites might be uncertain of the interdisciplinary approach found at the SFI and similar research centers. Ellen Goldberg, then president of the Institute and a former dean at the University of New Mexico, recalls the difficulties she had trying to set up a symposium at the university on chaos theory. People weren't willing to meet, Goldberg (SF) says. "Well, when I brought this to the research policy committee at UNM about this day-long symposium it was as if I asked people to give up their right arms. It was the most unbelievable response. People said—at least a couple of people on the research policy committee, 'Why should I walk over to the college of fine arts? By golly, I have to write my grants; I have to teach my students; I have to do this.'" The openness to interdisciplinary studies at the Santa Fe Institute, by contrast, is what attracted Goldberg. "You have to be a risk taker," she adds. Goldberg says that some scientists and academics may be unwilling to launch into interdisciplinary work because they fear incommensurability problems. "Who has time to learn this other language, the language of economics or physics?" she asks.

Indeed, reluctance to do interdisciplinary work often is a result of time limitations, or of fears that attention diverted from one's primary field will not offer enough monetary or career-advancement rewards to be worth the commitment. If a researcher in Goldberg's field of immunogenetics attends an interdisciplinary colloquium, for example, and sees strange equations on the board, he might wonder, "What does this have to do with my field?" That scholar may be concerned that he is wasting time that could be focused on work that will lead to a grant or tenure, Goldberg (SF) says. Such concern is amplified when granting agencies such as the National Institutes of Health (NIH) lack the expertise to review grant proposals that are outside the mainstream of a specific field.

Goldberg says that NIH and other agencies are slowly warming up to the idea of interdisciplinary research. For example, now it is acceptable to use the term "complexity" in a grant proposal, she says. "When NIH and other mainstream agencies and private businesses fund the Santa Fe Institute, others become interested, Goldberg says. Her comments echo Kuhnian paradigm theory, which maintains that the day-to-day activities of "normal science" respond slowly to major "paradigm shifts" in scientific understanding. Perhaps the Santa Fe Institute's interdisciplinary ethos is gaining acceptance as the foundation for a legitimate, normal science method. As many SFI members are quick to point out, reality is interdisciplinary; communication across disciplines, however difficult, occurs constantly. The very phrase "complexity science" suggests that reality does not fit tidily into discipline-specific paradigms.

Ballati (SF) argues that one of the Institute's greatest contributions to science may be in raising awareness that, "It's a messy world." Her work at the Institute has helped her affirm a kind of situational code of ethics, which requires human beings to consider their relationship to that world in all its facets, rather than with hard-and-fast rules, when making decisions. "I've discovered that it's not content as much as context," Ballati says. Problems like poverty, for example, "are big and they're messy and they cut across lots of disciplines." As we have seen, language also is messy.

Indeed, scientists at the Institute seem to share a reverence for the messiness of the world and for the forms of life that have evolved and organized themselves successfully within it. These scientists may be trying to understand how that ordered life came to exist. They even may be trying to replicate the process through computer simulations. But their pursuits do not suggest that emerging systems, especially living systems, are any less sacred because they are subjects of research.

During one visit to the Institute, I recall the entire courtyard area was closed off to allow an injured mountain blue bird time to recover and fly away. When I was interviewing Lachmann, he stopped mid-interview—actually mid-sentence—to gently capture a moth in his hand and carry it outside. A six-foot-plus-tall example of the most complex life form on earth stops to rescue an inch-long version of one of the simplest. It made perfect sense at the Santa Fe Institute. Yes, these scientists are beholden to metaphors and mathematical equations as they try to understand that world and develop their understandings for disparate audiences. But life in all forms and sizes, given by chance and the subtleties of such a complex world is what these Santa Fe Institute scientists revere.

6

"Complexity": An Etymology Leading to a Discussion of Whether it is a Metaphor or Something Else

Type the key word "complexity" into the search engine for any university library's electronic card catalogue and you will find page after page of holdings that contain the term, either in their titles or in the catalogue's abstract describing the article or book. The library at the University of Texas at El Paso, for example, returns 234 items to a search for this key word. The titles reveal just how far the concepts of complexity science have spread. As might be expected, there are plenty of books that consider complexity in the physical and mathematical sciences, with titles such as *The Emergence of Complexity in Math, Physics, Chemistry and Biology* (Pullman) and others such as *Cellular Automata and Complexity* (Wolfram). But you can also find books with the word "complexity" in their titles in the social sciences, business, even literature. Some books offer insights from the new science into matters of public policy and governance; others turn an explanatory theory into self-help applications to everyday problems in business; still others employ complexity theory to analyze the plots of great novels.

Scanning this list of books you quickly become aware of an etymological evolution—a bifurcation—in the use of the term. Books published before the mid 1980s typically use the word to suggest a state of being that is difficult to unravel, one that is imbricated with many layers of meaning. "Complexity" in older books has a colloquial meaning that is close to the way we use the word "complicated." So we might say, for example, that race relations in the United States or religious issues in the Middle East are complex, meaning that they are complicated by many human allegiances, emotions, superstitions, and historical memories. With few exceptions, these older works do not use "complexity" to refer to a specific theory. The exceptions are found in a few early works in information theory dating to the early and middle twentieth century. Some of these do discuss complexity in a way that is close to the SFI sense. A precursor theory known as "general systems theory" had emerged by mid century. Still, the concept did not flower across many disciplines as a meta-theory until that period beginning in the early 1980s—about the time the Institute was born in 1984.

As this chapter will show, the word "complexity" appears in science as far back as the early eighteenth century, although the most frequent occurrences are found later. In the arts and philosophy, the idea of complexity can be traced to Aristotle. The ideas that would give rise to a complexity science were taking shape in late nineteenth century and early twentieth century sciences of biology, psychology, thermodynamics, and statistical physics. Although none of these uses captures fully the notion of complexity that would emerge in information theory and, eventually, in postmodern SFI-type science, the various earlier uses have some attributes in common with what we would now consider the new science.

The difference between old scientific uses of "complexity" and the new usage is that complexity previously described specific behaviors or phenomena in reality. Now it embraces a new meta-science that isolates complexity as an overarching phenomenon to unite all of these member sciences. To use the language associated with Kuhn, we could say that a paradigm shift that gave birth to the formal science occurred in the late twentieth century. The paradigm shift came after many years of "normal science," where investigations in various fields isolated different kinds of behaviors that all somehow could be labeled "complex."

This word links researchers from many fields who suspect that interactions among agents and environments yield surprising results. "Complexity" functions for this postmodern science of affiliated disciplines because it accounts for various phenomena, but without being so restrictive in meaning as to suggest a dominating metadiscourse. Recall Erica Jen's comments to me when I was researching the "Initial Impressions" report for the Santa Fe Institute: "We do have a core set of metaphors that relate to complex adaptive systems," she said. As we have seen, when SFI researchers refer to complexity they are usually referring to the study of surprising order, or systems, which arise out of simple rules among agents that interact. If those systems are able to adapt to changes in their environment—if the agents can learn and adjust their fitness—then the complex systems are said also to be "adaptive."

Yet, even after the paradigm shift, the function and meanings of complexity and related terms remain elusive; we saw this elusiveness in meaning when scientists debated through their interviews the harmonic implications of many terms that relate to complexity science. Words that have metaphoric associations, such as "fitness landscape" and "game theory," present all kinds of insights and problems to scientists because they invite semantic discussions, which essentially are discussions about meaning. These issues of word meaning are central to scientific epistemologies. Little doubt could remain that discussions of individual words are important in creating knowledge at the Santa Fe Institute. If we could show "complexity" to be a metaphor, we could then argue that metaphor led to the paradigm shift which converted a general descriptive term into a specific new meta-science.

The goal of this chapter, therefore, is to take a close historical look at this central term and try to decipher its power. I will conclude that "complexity" is not a true metaphor, but that it functions as a collective abstraction that shepherds more concrete terms into making metaphoric contacts across fields. Because "complexity" is not a metaphor, it can describe a meta-science without conjuring up the troubling harmonic associations that are present in theory-constituting metaphors. By recognizing when abstractions like "complexity" are appropriate, and when more evocative concrete metaphors are preferred, scientists and those who write science can incubate and nurture a fragile new meta-discipline while still subjecting its component theories to the rigors of testing by semantic debate.

THE WORD "COMPLEX" IN THE *OXFORD ENGLISH DICTIONARY*

Invariably, any attempt to trace the history of a term in the English language leads to *The Oxford English Dictionary*—a multi-volumed work that offers pages of etymologies of words and traces their roots. SFI biologist Harold Morowitz has trodden this ground in search of help with the word "complexity." He briefly examined different definitions of the term in a 1996 article titled, "What's In a

Name? One Place to Look for 'Complexity' is in the Dictionary." As Morowitz found, the *OED* does not disappoint with reference to "complexity," providing four full pages of definitions and examples for the word in its various forms. These include the strange fellow noun "complexedness;" the adjective (also a noun), "complex," an adverb, "complexly," a verb, "complexify;" as well as the related terms "complexion" and "complicate." The two forms that seem most relevant to SFI science are the adjective "complex," as in "complex adaptive system," and the noun "complexity." These are the forms we will consider in detail.

"Complex" as an adjective derives from the Latin noun *complex—us*, which is a compound word, *com* meaning "together," and *plexus* meaning "plaited" (*OED* 613). So the term means "plaited together," which invokes images of fabric or material that is made up of different strands, but which takes on a new function and set of characteristics when woven together. The *OED* definition of this adjectival form continues, "Consisting of or comprehending various parts united or connected together; formed by combination of different elements; composite, compound. Said of things, ideas, etc." (613). Of note is the similarity to the term "complicate," which is often interchanged in everyday speech for "complexity." The Latin roots for "complicate" mean "to fold together" (615).

The entry for the noun "complexity" defines it tautologically as, "The quality or condition of being complex" (614). The *OED* then elaborates with three definitions:

1. "Composite nature or structure"

2. "Involved nature or structure; intricacy"

3. "An instance of complexity; a complicated condition; a complication" (614)

The *OED* offers various citations from works of science, philosophy, and literature that either contain some form of the word "complexity" or posit a reality that involves the interaction of simple components to form complex associations. The earliest references are in works related to philosophy, the arts, literature, and politics, showing clearly that the idea of "complexity" emerged in scholarship dealing with human society long before it took on discipline-specific scientific meanings. This is not surprising; science began as natural philosophy in ancient Greece and, as we have seen, did not depart from human narratives and myths until well into the Enlightenment. Many of the early uses of "complexity" place it in contrast to "simplicity," a dichotomy that Murray Gell-Mann has attempted to revive in his analysis of the Santa Fe Institute's lexicon. We will consider Gell-Mann's work later.

Aristotle's study of Greek narrative and drama, *Poetics*, discusses complexity in relation to dramatic plots. "Some plots are simple while others are complex . . .," Aristotle writes (19). A simple plot for Aristotle is one unified with continuous action, while a complex plot involves surprises, what we might now call plot "twists." Again, the notion of different levels of reality woven together is present.

Enlightenment philosopher and mathematician Gottfried Liebniz postulates a similar dichotomy in nature, proclaiming that reality is made up of substances capable of action that are either "simple" or "compound." Simple substances are "monads," a term that Liebniz used to describe the essential, atomistic states of being rather than actual atoms. These states include "lives, souls, spirits" (522). A compound substance is a "multitude." Liebniz argues that an animal or person is a compound substance made up of the essential substances, such as the creature's soul and spirit. In Leibniz's 1714 work, *The Principles of Nature and of Grace, Based on Reason*, it would seem that the word "complex" could be substituted for "compound," in the Latin sense, whereby a living creature is the result of many essential qualities that are plaited or woven together. Leibniz foreshadows a Santa Fe Institute type of complexity when he argues that monads interact in such a way that "everything is connected and each body acts upon every other body, more or less, according to the distance, and by reaction is itself affected thereby . . . "(523). Yet, the monads do not evolve and, presumably, would not adapt to changes in their environment.

John Locke, the English philosopher of the early Enlightenment, is best known for *An Essay Concerning Human Understanding*, a treatise on epistemology that also posits a dichotomy between simple and complex ideas. Published in 1689, this work is anti-Platonic in that it rejects the notion that knowledge comes about by looking inward to discover innate essences of reality. Locke triumphs a new scientific philosophy of empiricism, looking outward, although he argues that human understanding occurs when such insights are processed rationally. Simple ideas are what we would know as sensations, such as the sensation of cold, heat, light, darkness, motion, and so forth (132). The mind is a passive receptor of such simple sensations, Locke argues. However, when the mind processes simple ideas collectively, it creates complex ideas. So, Locke argues, awareness of color and shape may lead a person to proclaim a particular setting to be beautiful (165). Beauty and other abstractions are complex ideas; this argument conjures up the image of plaiting together basic elements to form something greater.

Presumably, this type of complexity would be similar to the Santa Institute's use of the term in that the all-embracing beauty of a landscape, for example, is more than the sum of the individual aspects of that landscape—sunlight, terrain, cloud formations, etc. One could argue as an extreme empiricist that each element changes the other, as when the sunlight falls on the hills and crevices of a terrain in ways that makes that terrain appear to be different at different times of day. Yet, it would be a stretch to suggest that the landscape itself "adapts" to each component.

Other references to complexity during the Enlightenment are found in works of political philosophy by the English utilitarian Jeremy Bentham and the Irish politician Edmund Burke. Bentham's 1789 treatise on ethics and law, *An Introduction to the Principles of Morals and Legislation*, categorizes offenses as being "simple" or "complex" (243). An offense is simple if it hurts only one aspect of an individual's life, say her reputation; it is complex if it damages several aspects, say her reputation and her property. This use of the term "complex" is more additive than interactive and, hence, draws little comparison to the later Santa Fe Institute usage. Burke, in his 1790 *Reflections on the French Revolution*, suggests that human society is complex, although he does not draw attention to the word by defining it, and nor does he contrast it to notions of simplicity. "The objects of society are of the greatest possible complexity," Burke writes (quoted in the *OED*, "complexity" 614).

These first use of complexity in the sciences, at least according to the *OED*, is in early Enlightenment astronomy. That astronomy would be the first site is not surprising, given that science began in the contemplation of the heavens. Morowitz's exegesis of this usage notes that astronomer John Keil in 1721 translated the work of Pierre Louise Moreau de Maupertuis, who considered the complexity of celestial motion ("What's in a Name?" 7). The notion of complexity emerged in mathematics at around the same time, when mathematicians found the need for complex numbers to solve certain quadratic equations whose solutions require the square root of a negative number.

Another scientific reference that Morowitz researched is attributed to the American essayist, Ralph Waldo Emerson, whose 1847 lecture on Johann Wolfgang von Goethe considers structure in science. As Morowitz quotes Emerson, "It is the last lesson of modern science that the highest simplicity of structure is provided, not by a few elements, but by the highest complexity" ("What's in a Name?" 8). According to Morowitz, Emerson is referring to Goethe the botanist and morphologist rather than Goethe the poet. Morowitz notes that Goethe "might be saying that the one emerges from the many; that unity arises out of diversity—an idea found in modern complexity theory" (8). Also of interest is the rhetorical antithesis in Emerson's sentence whereby the reader is surprised to find that what would seem to be predicated by "a few elements" is not "simplicity," but is instead "complexity." The element of surprise that emerges from the seemingly antithetical relationship between the simple and the complex is integral to later Santa Fe Institute uses of the term "complexity."

The noun form of the word "complex" is most commonly associated with psychology—a usage that came into vogue in the nineteenth century. Such usage is defined by the *OED* to mean "a group of emotionally charged ideas or mental factors, unconsciously associated by the individual with a particular subject, arising from repressed instincts, fears, or desires, and often resulting in mental abnormality . . . " (613). Notions of an "inferiority complex," a "Napoleon complex," and similar personality tics are commonplace in popular psychology. Sigmund Freud made common the idea of a mother fixation, known widely as an "Oedipus complex."

The term entered the literature of psychology in 1907 through the work of Carl Jung, although it originated a year earlier by another psychology theorist (*OED*, 613). For Jung, a complex was a "cluster or constellation" of contents that lay in the personal unconscious (Hall and Nordby 36). Jung found in research that a person sometimes delayed responding to certain words because those words conjured up associations with other memories held by that person. This understanding of "complex" suggests that interactions among ideas lead the mind to a more powerful idea, or complex, which is greater than the sum of the individual ideas. So Jung's definition of "complex" is strikingly similar to the Santa Fe Institute usage, although it would be impossible to know what rules, if any, the human mind followed in making memory associations. Also of interest, a Jungian complex involves endless wordplay and associations, which are the very phenomena that allow metaphor harmonics to spawn new insights. The act of using metaphors, then, could be said to be complex thinking in the SFI sense of the word.

The nineteenth century also saw gradual adoption of the word "complex" and its cousins in the biological sciences. The naturalist Charles Darwin referred to nature's complexity in an 1859 letter to geologist Charles Lyell, who had just penned some comments about Darwin's newly published theory of evolution. Lyell was known for his theories of gradualism, suggesting that geological and biological changes occur slowly and evenly. New and more advanced species would arise to take the place of older ones only after their tenure had ended, perhaps in keeping with the inevitable plan of a divine creator to bring forth the ultimate species—human beings.

For Darwin, evolution was not so peaceful and harmonious, but instead was a battlefield in which all sorts of accidental variations of life would emerge and struggle to survive. Hence, Darwin responds to Lyell's suggestion that the forms taken by living creatures improve gradually throughout history. (Lyell uses Leibniz's term "monads" for these forms.) "I grant there will generally be a tendency to advance in complexity of organisation, though in beings fitted for very simple conditions it would be slight and slow," Darwin writes in letter to Lyell. "How could a complex organisation profit a monad? if [sic] it did not profit it there would be no advance" (Darwin, Francis 6).

Darwin's implied argument is that complexity is not the goal of evolution, for evolution has no goal. He does not define "complexity," but aligns it with the idea of organization, which foreshadows the SFI usage. A similar usage in the nineteenth century appeared in a zoology text attributed by the *OED* to an author, Blake, which refers to the stomach as a complex organ (613). It bears notice that complexity sciences at the Santa Fe Institute retain this organic, biological emphasis, albeit now focused more on postmodern artificial life systems than on real ones.

"Complex" appears in the literature of organic biochemistry and inorganic chemistry. According to the *OED*, it refers generally to "a substance formed by the combination of simpler substances, especially one in which the bonds between the substances are weaker than or of a different character from those between the constituents of each substance" (612–613). The *OED* carries a citation apparently from a dictionary of science or biology in 1895 that defines cellulose as "a fur-fural-yielding complex, which appears to be an oxycellulose deriva-tive" (613). "Complex" also can be a verb in chemistry, according to the *OED*, as in this 1960 citation from an Australian science journal: "Tannic acid was also found to be capable of complexing small amounts of copper" (613).

"Complex" as a noun has a more specific usage in inorganic chemistry, where it is synonymous with the term "coordination com-pound" (Hampel 69). Such a complex is a molecule "formed by the attachment of a transition-metal to another molecule or ion by means of a coordinate covalent bond . . ." (69). Basolo and Pearson's 1968 study of coordination compounds sets out to explore the mechanics, or steps taken, as atoms, radicals, and ions react with one another. "Such compounds contain a central atom or ion, usually a metal, and a cluster of ions or molecules surrounding it," they write (1). Basolo and Pearson note that the "historical development of the chemistry of coordination compounds dates back approximately to the end of the eighteenth century" (1). If so, the study of chemical complexes would represent one of the first applications of a notion of "complex" to a specific discipline—pre-dating its specific use in psychology. Here we again find the notion of many parts interacting, although as Fontana (SF) has pointed out, chemistry has tended to treat the reacting enti-ties as variables that take a certain value in a function (reaction), rather than as functions themselves.

Chemists and physicists went through their own paradigm shift in the late nineteenth century and uncovered new insights about the strange behavior of matter, which made possible the rise of complexity science 100 years later. According to historian of modern thought John Herman Randall Jr., chemistry and physics merged "in their roots" in the nineteenth century when both recognized that the com-plex behavior of the atom lay at the foundation of material reality. "Chemists, bringing order into their science by a verifiable atomic theory set in mathematical terms, discovered the Periodic Law of atomic weights, and were led to the same analysis of the atom into electrons and a nucleus of varying complexity which had been neces-sary in physics," Randall writes (470).

Enlightenment physics had been defined by Isaac Newton's laws of motion, known as "Newtonian mechanics." This science studied the motion of relatively large bodies, such as planets, and postulated how various forces and properties affecting those bodies would govern their motion. The key to Newtonian mechanics is that each body is considered individually; equations governing motion derived values for each body. In the latter half of the nineteenth century, Austrian physicist Ludwig Boltzmann founded a new branch of physics that looked at the collective behavior of microscopic-sized bodies. Attention shifted from the individual to the group by using statistical mathematics techniques to calculate probabilities that the particles would occupy certain states, i.e., that they would have an exact position and momentum. So instead of the Newtonian certainty of large-body behavior, this new science of statistical mechanics dealt in probabilities of particle distribution.

Boltzmann employed statistical analysis of particle behavior to tackle problems in thermodynamics, the relatively new science of heat and energy. The second law of thermodynamics is the law of entropy, which holds that without some external influence, energy dissipates. A cup of coffee cools. Order—in this case water molecules holding heat energy—dissolves into disorder as entropy increases. That heat can never be recovered, which is the reason that physicists see time as an arrow that always flows in one direction. Boltzmann postulated that maximum order is present when a quantity of molecules in a container are all in the same arrangement, which in our example would mean that all the heat in a region is concentrated in the coffee cup.

This condition of concentrated heat is improbable in the universe, except in the rare cases when there is someone to turn on a stove. For that matter, the sun and other stars are improbable concentrations of heat. In the more probable outcome, molecules can be anywhere; they can occupy an infinite number of arrangements. Boltzmann used the term "complexion"—a variation on "complexity"—to mean an arrangement, or state of being (Capra 187). More "complexions" means greater entropy, which is the most likely outcome (188).

The *OED* reveals that the term "complexion" as used in nineteenth century thermodynamics carries almost the same meaning as it did in the philosophy and sciences of the Middle Ages. In that time, "complexion" meant the "combination of supposed qualities (*cold* or *hot* and *moist* or *dry*) in a certain proportion, determining the nature of a body, plant, etc . . ." (613, italics original). We use "complexion" in everyday speech to refer to the particular color characteristics of a person's face, again drawing attention to qualities or arrangement of being.

Of course, "complexion" in the thermodynamic sense relates to a statistical concept, whereby an environment is seen to be complex if it is random. It is complex if it cannot be reduced to description (or "compressed," the word used in information theory) because it literally offers no pattern from which to distinguish it from ambient background. As science writer John Horgan notes, "according to this criterion, a text created by a team of typing monkeys is more complex—because it is more random and therefore less compressible—than *Finnegans Wake*" (197).

This sense of "complexity" appears to run counter to the SFI sense of order emerging among interacting parts. SFI scientists are aware of this problem of overlapping associations with their key term. Shalizi and Crutchfield in their SFI paper on patterns in physics, which we encountered earlier, note that complexity in a statistical sense (known as "computational complexity") may be ill suited for describing natural systems because it is "a failure to capture structure" (4). Gell-Mann, in his book *The Quark and the Jaguar*, substitutes the term "algorithmic information content" for the notion of computational complexity. He contrasts a system that is random and cannot be described with systems that are so ordered that they can be easily described. (For instance, "All the heat is in the cup.") For Gell-Mann and other Institute members, the most interesting complexity occurs between the two extremes, when there is some degree of unpredictability, but the system still yields patterns (59).

We recall that the knowledge of entropy was integral to the mid-twentieth century development of information science, "cybernetics," which asserts that the attempt to send a signal over a wire is always frustrated by the tendency of the system to swallow that signal in noise. Claude Shannon used the term "information" when he might have been better off using "signal;" his use of "information" led to the same semantic confusion that has arisen out of the concept of "computational complexity." As Capra (64) shows, the word "information" implies meaning. In this sense, the information flowing over the wire orders the world for a receiver by providing an intelligible message. The way Shannon used "information," however, suggests an amount of available data in a signal, even when that data may be meaningless. So in Shannon's sense, we may learn that ten people flipped a coin and four got heads, while six got tails. We would have ten bits of "information," but it would not mean much. If we learned, however, that the leaders of India flipped coins and decided to bomb Pakistan, we would have just two bits of information, but it would be much more meaningful.

This debate remains in science over whether "complex" means something rich with data (even if it is meaningless and random to the human observer), or whether it means something less crowded but more likely to yield interesting patterns. Essentially, our understanding of what is "complex" in part depends on whether we are considering biotic (i.e., living) systems, where the potential for meaning and patterns discernible by an organism is important, or whether we are describing abiotic (i.e., physical or mechanical) systems, where the number of statistically possible arrangements of reality is what is important.

Russian philosopher, economist, and political revolutionary Alexander Bogdanov in the early 1910s postulated the idea of organized systems in biology, capturing very much the sense that meaningful complexity is when agents adapt to information—the sense most often considered at the Santa Fe Institute. According to physicist Fritjof Capra, Bogdanov's theory of "tektology" describes mainly biotic and social systems where the whole is greater than the sum of the parts. Arguably, then, Bogdanov could be seen as the founder of postmodern complexity science. He uses the term "systems" and "complexity" interchangeably (Capra 44). Austrian-Canadian biologist Ludwig von Bertalanffy spread these ideas between the 1940s and the 1960s under a holistic science known as "general system(s) theory." Bertalanffy's treatise, *General System Theory*, argues that society and its technologies function "as a tremendously complex network of interactions" (4). The argument begun here and carried on at the SFI is that these exchanges of information among a group of agents are what give rise to reality as we know it.

Complexity theory of information processing from a cognitive psychology perspective appeared in the 1950s and 1960s. *Complexity, Managers, and Organizations* by Siegfried Streufert and Robert W. Swezey, published in 1986 as part of a series on organizational and occupational psychology, describes the historical development of early complexity theory in cognitive psychology. Information available to an individual or an organization is processed according to "a bipolar cognitive scale," a type of binary thinking known in this field as a "dimension" (16). The usage here recalls the emphasis on dimensions in mathematical physics that arose throughout the twentieth century.

In this case, for example, a business manager may see the world in terms of profit and loss, or productivity and inefficiency. These dimensions (we could also use the term "schemata," in keeping with cognitive theorists) determine how we perceive new information.

A cognitively complex person is one who is able to apply many dimensions to reality, one who is not restricted to just a few binaries. So instead of seeing all problems as those of profit and loss, a complex manager would also take into account longterm strategies, employee needs, and other factors. These theories of organizational science argue for an approach to organizational decisionmaking that is adaptive. A cognitively complex person in an organization is adaptable enough to apply creative solutions to problems, instead of merely defaulting to the established binaries (74). As we might expect, the highest level of complex thinking occurs not with the maximum amount of information coming in, but at an intermediate level. This theory is consistent with the common understanding that too much information—information overload—is stressful and counterproductive. Again, we find that a little bit of meaningful data is better than a lot of noise for inspiring complex adaptations.

"COMPLEXITY" ARRIVES IN SANTA FE

We can see without struggling further with the brain twisting semantics of entropy and information theory that the term "complexity" in its various forms arrived at the Santa Fe Institute in 1984 pregnant with powerful but illusive scientific meanings and harmonics. Clearly those harmonics have not diminished with the paradigm shift that established a new branch of interdisciplinary complexity science. Members of the Institute openly point out that complexity is ill defined. In the early 1990s, physicist Seth Lloyd, who is associated with the Institute, proposed a list of some thirty different scientific definitions now applicable to the term (Horgan 197). While using a significant portion of his book to consider the meaning of complexity, Gell-Mann acknowledges in *The Quark and the Jaguar* that, "Any definition of complexity is necessarily context-dependent, even subjective" (33). For Horgan, these definition problems "highlight the awkward fact that complexity exists, in some murky sense, in the eye of the beholder (like pornography, for instance)" (197). Yet, scientists have trouble sitting still for assertions of rank subjectivity. Gell-Mann asserts that while complexity need not have a precise definition, the presence of the term in Santa Fe Institute-type science "implies the belief that any such system possesses at least a certain minimum level of complexity, suitably defined" (27).

Arrow (SF) in his interview suggests that scientists following the positivist tradition indeed bristle at a term like "complexity." It cannot be easily pinned down or falsified because it is not isolating reality and making a positive claim about it. "I think complexity . . . is a negative word," he says. "It's the absence of simplicity. . . . And, therefore, the negation of something is usually rather vague." Cowan (SF) says, "Complexity is a funny word," adding that the term has heuristic value, but that calling systems complex "is not terribly useful." Shalizi (SF) says that despite all the talk at the Institute of finding a measure of complexity that can apply across different systems, "I don't think that we're going to find any one thing which complexity is good for."

But even given this impatience with a vague term, our interviewees have tended to agree that much of the power of descriptive words, including metaphoric words, lies in their flexibility to change connotations according to situation. These chameleon-like terms allow scientists to see new associations among seemingly disparate ideas, which is the crux of the Santa Fe Institute methodology. Struggling to find rigorous definitions, however frustrating that process may be, is as much a part of science as is laboratory field work. Institute scientists have not shied away from this lexical research; nearly all the books that SFI scientists have written include discussions about the meaning of complexity.

Take, for instance, Gell-Mann's linguistic treatment of the term in his book. Like Arrow, he starts by contrasting complexity with simplicity, defining the latter loosely as "the absence (or near-absence) of complexity" (27). Gell-Mann then parses the term "simple" to its Latin and Greek root to mean "once-folded," where sem is "one" and plicare is "folded" (*Quark* 27, *OED* 495). "Note that both "plic-" for fold and "plex-" for braid come from the Indo-European root "*plek*" (*Quark* 27). Gell-Mann—a scholar of linguistics as well as of the sciences—has suggested that Institute science be known as "plectics," a variation on the root word. This never caught on, however; we know that language often does not conform to prescriptions for change, even when those prescriptions might seem logical. The proposed new language must also resonate within the ear.

So here we have the metaphoric root of the concept of "complexity," whereby the level of complexity a system exhibits is determined by how many folds it contains. Still, reverting to the root meaning does not solve fundamental definitional problems, such as whether complexity requires a lot of information, or less information but with greater meaning. One of Gell-Mann's most illuminating moments in his *The Quark and the Jaguar* is when he looks at some of the contradictions in the different kinds of complexity. "Computational complexity has proved to be quite a useful notion," Gell-Mann writes, "but it does not correspond very closely to what we usually mean when we employ the word complex, as in a highly complex story plot or organizational structure" (28).

Gell-Mann notes the problems that ecologists have encountered in trying to define complexity in a tropical forest. An elementary notion of a complex forest could be derived by counting the number of trees, birds, insects, etc. But ecologists also would have to consider interactions among different organisms in the forest—which bird eats which insect, and so on. A definition of complexity in this instance would depend on whether one considered raw data (number of species) or the stories told by those data (interactions of species) or both measures, as well as the level of detail one chose to focus on when considering the forest. For example, as Gell-Mann writes, in studying the complexity of a forest you would have to decide whether to consider just large organisms, such as animals and birds, or whether also to take into account smaller organisms, even to the level of viruses. When physicists choose to ignore some detail they are said to be "coarse graining." Gell-Mann writes, "Hence when defining complexity it is always necessary to specify the level of detail up to which the system is being described, with finer details being ignored" (29).

Gell-Mann, then, is attempting to resolve semantic problems with the term "complexity" by viewing reality as a field of vision, such as that visible through the eyepiece of a microscope. The scientist determines *a priori* what will constitute complexity, just as he does in deciding what power lens to use on the microscope, which determines the level of detail he will see. This recalls metaphor theorist Max Black, who argues that metaphor works because a literal term acts as a "frame" that surrounds a field of vision whose "focus" changes as various metaphoric associations wash over the picture. Gell-Mann, of course, is arguing that scientists as observers literally must decide the level of specificity to consider, while Black suggests that metaphor brings different pictures into focus, and does not merely amplify an existing one. Still, it is interesting to note the similarity in the concepts.

To further resolve meaning, Gell-Mann dissembles the components of a system that involves information and, in the process, adds the modifier "effective" to the word "complexity." He distinguishes information from meaning by borrowing a term from computer science, "algorithmic information content." A string of numbers 1101l0110110 . . . repeated for a mile, let us say, would be full of information, but it would be simple at its essence (38). It could be "compressed" to a rule or a computer program that says, "Repeat 110 for a mile." The length of that program is the "algorithmic information content" contained in the string of numbers; in this case it is a pretty short program. By contrast, a truly random set of numbers repeated for a mile could not be compressed at all. The only way to describe that set would be to write out the whole mile's worth of integers. It would have much higher algorithmic information content than the first example, even though both contain the same amount of data.

Neither would be very interesting to most SFI complexity scientists, however, because the first example would be reducible to a trivial rule (repeat 110 for a mile) and the second could not be explained by any rule. What would be interesting, by contrast, is the number π in mathematics. It is an infinite string of integers (lots of information), and it can be compressed to a rule (divide the circumference of a circle by its diameter) giving it an intermediate level of algorithmic information content. But the compression is not trivial, in part because it involves concepts of circumference, diameter, and ratio. The rule is not immediately obvious looking at the string of numbers in π (3.1415 . . .), but the when the rule is discovered it reveals a pattern of relationships that is ubiquitous in geometry. In other words, a simple but non-trivial rule defines a lot of patterns that shape reality. Gell-Mann refers to this middle level of algorithmic information content, where a simple rule explains a lot, as "effective complexity" (58).

Gell-Mann has resolved some of the definitional problems of the key term, "complexity," by modifying it with the adjective "effective" to add specificity. By modifying a word after encountering its contradictions, Gell-Mann shows how problems of semantics can lead to new scientific insights. Other SFI scientists have used this modifying approach to lend specificity to the term. In their writing, John Casti and John Holland refer more often to "complex adaptive systems" rather than to the idea of "complexity" on its own. Casti, in his book, *Would-Be Worlds*, takes several pages to tease out what he calls "a viable theory of complex, adaptable systems" (213–214). Such a system, Casti writes, has a medium-sized number of agents, (1) that are intelligent and adaptive, and (2) have local information, meaning that they interact with a small number of their fellow agents (3). These are the components of almost all complex systems, Casti says (214).

Note that the second component of Casti's definition would seem to restrict the focus of complexity to systems made up of living (or artificially living) organisms. The key here is that these organisms must be capable of learning and of modifying or adapting their behavior according to that knowledge gained. The living organisms interact with each other, but not with so many fellow members that the information dissembles into noise. Those organisms then change behavior according to their interactions. For Holland, the notion of "adaptation" is crucial to his systems. He writes, "Adaptation, in biological usage, is the process whereby an organism fits itself to its environment" (9). Of course, the environment is made up of other adaptive agents, so each is responding to the other, leading to overall changes in the system.

Again, we find ourselves faced with the concept of "living," which we know is not a trivial concept in science. Presumably, a nonliving system could be complex, but it would not be adaptive. Recall Cowan's example of three large bodies held in close proximity in space by their gravity. They would be a medium-sized number of agents that interact with each other via gravity. Of course, they would not "learn" anything from that interaction; they would not adapt. But the mathematical description of the physical forces governing their motion, based on those interactions, would be highly complex, in the sense that the calculation would require many twists and turns—it would be "folded over" many times. The rule that would describe the behavior of these three bodies would fall in the middle range of Gell-Mann's algorithmic information content; it would not be trivially short, because the mathematics is difficult, but it would not be so long as to simply state the position of every body at every point in time.

But if we attempt to back into a definition of "complexity" by stripping out the adaptive part of Casti's definition, we are not left with much of discriminatory value. "Complex" would simply mean a pool of agents that interact. Pebbles lying next to each other in the desert could be "complex" because they form a group whose members exert a tiny gravitational and perhaps electro-magnetic force on each other. But these forces are too small to influence each pebble's position. Certainly no one would seriously consider this group of pebbles to be complex.

How about the earth and moon as a system? Is it complex? Is it adaptive? Certainly these are bodies that interact. They also cause changes in one another, as the power of the tides reveals. Of course, the earth is not "learning" from the moon, but it is changing cyclically every day. Here we are approaching Casti's second criteria—that of adaptation—and we are coming closer to an interesting, complex system than with the pebbles. But because the tidal patterns presumably are a determined function of the relationship, no learning occurs. Therefore, we are led back to the notion of non-linearity, indeterminacy. A complex adaptive system must be rule-governed, but for Casti and others at the Institute, there also must be the potential for indeterminate surprises.

If we consider also the wind and other global weather forces along with the moon, we find that there often are surprises in tidal flows; these surprises sometimes lead to coastal flooding. The tides, then, are governed by complex interactions. These interactions also are subject to the rules of chaos theory, in that a small change in any one factor—say a tiny change in the barometric pressure over the Pacific Ocean—could have profound effects. While the forces that combine to cause the tides could not be seen as adaptive, perhaps they could be seen as "surprise-inducing" because they result both from the laws of physics and from chance.

We could go on for many more pages positing examples and putting them to the tests that Gell-Mann, Casti, and others have established for the term "complexity." To some degree we would be chasing our tails in trying to pin down a definition that works in all cases. As each example builds upon the example, we would find ourselves adding corollaries to the definition, and new modifiers. At some point, however, a skeptic would be certain to ask if in constantly modifying the term, we are changing it such that it no longer has much in common with other versions. Too many modifiers would eventually lead to noise, drowning out the powerful meaning of "complexity" that now sounds across a variety of disciplines. Too much specificity, as Bak and other SFI scientists have pointed out, means that scientists would lose the umbrella concept that forms the Institute; they would be driven back into their isolated disciplinary corners.

Gell-Mann is right to point out that any notion of complexity is subjective. Yet, such subjectivity need not ruin the value of the term. As we have seen, vagueness in terminology invites associations and stimulates scientists to explore whether their small view of the world might be similar or different from other views. Hence, several scientists offer deliberately vague definitions of the term. Holland in *Hidden Order* carefully explains seven basic properties and mechanisms he sees as intrinsic to a complex adaptive system. But he also provides a coarse-grained picture; for Holland, all complex adaptive systems exhibit "coherence in the face of change" (4). The stock market, for example, fluctuates daily. It also undergoes significant longterm changes, as we see now in the shift from a manufacturing to an information economy. Yet, the stock market retains a form today that would be easily recognizable to Adam Smith.

Stuart Kauffman offers a similarly pithy definition of complexity, capturing the same idea as Holland. For Kauffman, complexity is present whenever a system exhibits both "flexibility" and "stability" (*At Home* 87). Kauffman's definition for a complex system applies to the word "complex" itself. The word is stable enough to signify something intrinsic to various sciences, but flexible enough to embrace the distinctions that emerge from those sciences. The word "complexity" was complex enough to corral various ideas emerging in normal science over the past century and turn them into a more far-reaching paradigm shift.

THE NEW "COMPLEXITY" RETURNS TO OLD HAUNTS

The paradigm shift that has given birth to complexity theory has moved beyond science even as scientists debate the value of the new theory and its meaning. Notions of complexity again are found in the arts and social sciences, as they have been since Aristotle. But the term has returned to its old haunts echoing with the new meaning of the paradigm shift. A similar phenomenon occurs in popular culture. For example, the popular music paradigm shift of the 1960s occurred when styles found in American blues and rockabilly were exported to Britain, converted into a new rock 'n' roll sound, and sent back to the United States as a new sound.

The sense of "complexity" as used by the general public before the shift is found throughout the University of Texas at El Paso library catalogue. Take, for instance, an imprint of a 1978 speech by a U.S. State Department policy expert, titled *Managing Complexity in U.S. Foreign Policy*, which describes U.S. foreign policy concerns as "complex" (Lake). These policy concerns are labeled as such because they involve many facets—considerations of the arms race, world trade, hostile nations, and so forth—not because they display scientific properties of a complex adaptive system. A book from the late 1960s titled *Democracy and Complexity: Who Governs the Governors?* argues that our "mass society" is complex and difficult to manage because it is highly populated, highly industrialized, and highly urban (Krinsky). While U.S. foreign policy challenges and domestic governance issues certainly may be complex in the Santa Fe Institute sense of the word, this sense could not have been intended or conveyed in a speech or a book prior to the paradigm shift.

For contrast, consider the 1994 publication, *Managing Chaos and Complexity in Government: A New Paradigm for Managing Change, Innovation, and Organizational Renewal*. Here a government professor includes diagrams of strange attractors, bifurcation diagrams, and other images from the new fields of chaos and complexity theory to explain matters of public policy (Kiel). British sociologist David Byrne in his 1998 book, *Complexity Theory and the Social Sciences: An Introduction*, explores how the post-paradigm-shift theory can be used to analyze problems of public education, public health, and municipal governance. He cites Kauffman, Gell-Mann, and others at the Institute and freely uses metaphors such as "strange attractor," "non-linear systems," "fitness landscapes" and the like that are associated with the new lexicon.

So the distribution of public schools in the United Kingdom is seen as a fitness landscape. A feedback system emerges, for example, when parents refuse to send good students to schools in poor areas, thereby further degrading their quality. The idea that deficiencies in a school lead to a bad reputation, which lead to further deficiencies is not especially remarkable, but is made to seem more so by being cast in the new terminology.

Now look again back before the paradigm shift to find Robert J. Barth's 1972 book of literary criticism, which examines religion in the work of novelist William Faulkner. Barth could not use "complexity" in the new paradigm sense, although the library catalogue abstract refers to the "theological complexity of Faulkner's fiction." Here "complexity" means, among other things, Faulkner's complicated, twisted writing style and the multiple level of human emotions and religious beliefs that play out in his stories set in the American South—a region that most would see as complex in the sense that it embraces many conflicting social and religious values.

Compare the use of "complexity" in this abstract of the Faulkner study to Thomas Jackson Rice's analysis of the work of James Joyce, which was published 25 years later. Rice's 1997 book, *Joyce, Chaos, and Complexity*, has everything to do with the new interdisciplinary science. It includes images from chaos and complexity theories, including bifurcation diagrams and pictures of Mandelbrot sets of strange patterns, to explore plot turns in the novel *Ulysses*. "As Ulysses becomes increasingly chaotic, it often does so by an expansive feeding back into the text of elements already present in early 'states' of its text," Rice says (102–103). Another passage evaluating *A Portrait of the Artist as a Young Man* refers to "Joyce's sophisticated use of Riemannian geometry for the form of 'Ithaca' " (69).

The new paradigm language perhaps is most evident in the business-book press, a phenomenon that various Institute members who work with corporate sponsors have pointed out in their interviews. We find the theory converted directly to application in a 1999 publication by Susanne Kelly and Mary Ann Allison. Titled *The Complexity Advantage: How the Science of Complexity Can Help Your Business Achieve Peak Performance*, this book looks at self-organization among businesses and considers the economy as a web—a common post-shift metaphor. Kelly, a vice president at Citibank, writes briefly about her chief executive officer, John Reed, and his involvement as a Santa Fe Institute benefactor. This book freely employs the rhetorical strategy of dissociation, which we encountered in Perelman and Olbrechts-Tyteca's theories. The new information economy is dissociated from the old manufacturing economy in a chart. The new economy, for example, is knowledge-based, global, complex, and innovative, while the old economy is commodity-based, regional, linear, and repetitive (6). Much ebullience about the potential for the Internet to change the world, for example, has been synchronous with the paradigm shift toward complexity in the sciences.

The term "complexity" has come full circle. It first flourished as a descriptor in the arts, philosophy, and social-political sciences. It then evolved into a discipline-specific term among statisticians and information theorists, leading ultimately to a trans-disciplinary meta-theory at the Santa Fe Institute. After undergoing this transformation, "complexity" returned to the arts and popular culture as a means of interpreting literary plots, business trends, and other phenomena. These interpretations would not have been available before the paradigm shift raised complexity to the level of meta-theory.

IS "COMPLEXITY" A METAPHOR?

The paradigm shift occurred relatively suddenly. To borrow the SFI's sand pile catastrophe metaphor, which often crops up to describe incidence of earthquakes and other irregular natural and social events, limited uses of the term "complexity" in specific disciplines built up gradually—perhaps over several centuries. Eventually the pile of meanings cascaded over into a new meta-science. This word evolved from its loose association with the complicated and multi-faceted to a specific association with a new type of science that embraces these concepts, but also adds new associations of emerging order and mathematical regularity. The science that occupies Santa Fe Institute researchers exists because an old word gained new meanings. There could be no more convincing evidence of the power of language to shape knowledge than the realization that an etymological change accompanied a new science.

The movement of the word "complexity" from philosophy and the arts to science and back again in a different form seems suspiciously metaphoric, especially when we recall the root definition of metaphor as the process by which meaning is transferred across terms. Recall that most constructivist theories proclaim metaphor to be a device for producing knowledge. They have at their heart some sense that metaphor does its work by transferring ideas. If we say, as in Chapter 3, "the sun is a furnace," we come to see the sun not merely as a distant source of heat and light, but also close up as something that is capable of forging material change on the earth. We have transferred the notion of a furnace to that of the sun and also the notion of the sun (never ceasing power, primeval) to that of a furnace. This two-way metaphoric process is known variously as "interanimation" or "interaction" by constructivists.

But movement within a single term is not enough to make it a metaphor. To declare that "complexity" is a metaphor, we would have to ask, of course, what two terms form the tenor and vehicle, the frame and focus? In other words, what would "complexity" be a metaphor for, what would be the more literal term (assuming the concept of "literalness" has merit) and what would be the more figurative term? Certainly, if "complexity" is a metaphor, it is not as simple a metaphor as our example from Chapter 3. It is not immediately obvious how we would fill in the blank, "Complexity is_____.

We know of course that the presence of metaphor is not always obvious. A comparative term can be hidden. If a scientist speaks of an electron's orbit, she most likely will not clarify her statement to say that because an atom is like the solar system, an electron orbits like a planet. The metaphor is occluded; the very act of creating a metaphor involves fusing together the parts of an analogy so that the analogy is hidden. We could argue that "complexity" is a collapsed metaphor; the comparison is implied: "Reality is complex." In this sense, everything we know about reality would be colored by the concept of "complexity."

This argument seems hollow, however, because it stretches the concept of metaphor to cover any predicate relationship, thereby committing the big rhetorician's sin of subsuming everything under a single theory of knowledge. To proclaim that noun X is adjective Y (or has the qualities of Y) is one of the most common assertions of reality, a way of distinguishing relationships among pieces of that reality by describing the properties that each piece manifests. If we say, for example, that "Joe is happy," the concept of happiness washes over onto our image of Joe, in a stative or equative clause structure, and our knowledge of Joe as a person helps to give specific meaning to the abstract term "happiness." In that sense, all words standing in proximity to other words interact. Meaning is dependent upon context. But to make "complexity" a metaphor this way is tantamount to saying that all language is metaphoric, which may be true, but is not helpful in determining what it is about some metaphors that make them special.

We could also argue that "complexity" has held various meanings in different fields of thought over the ages, and that the term is a metaphor for all of those meanings. So if we refer to complexity in science, we immediately, but perhaps unconsciously, conjure up images of complexity in Aristotle's theory of drama, Jung's theory of psychology, and so on. In this sense, "complexity" is a metaphor for itself. Again, this would not be a useful insight, except to show that all words can be unpacked to discover a host of old buried meanings. Language, like society, is built upon the ruins of earlier societies. True enough, but what is special about metaphor?

The problem with calling "complexity" a metaphor is that the term is now an abstraction. I would argue strongly that metaphors require at least one concrete term to function. Their power is in putting flesh on concepts that are otherwise vague. If we say, "Love is sorrow," for example, we are predicating that the abstract condition of love has attributes that equate it with the abstract condition of sorrow. But neither abstraction offers metaphoric focus to the other, to use Black's terminology, because neither can turn the hazy images of the abstract into something concrete. As Lakoff and Johnson reveal, the sentence, "I am in love" is a hidden metaphor because it conjures up images of love as a container that holds an individual in captivity. Likewise, if we say, "Love is tears," we are making a metaphoric transference from the concrete image of someone crying to the abstract idea of love. The word becomes flesh. No such transformation occurs when two abstractions are placed in apposition.

"Complexity" is not a metaphor primarily because it is too abstract. It contains nothing concrete, as "fitness landscape," "signaling," "game theory," "spin glass," and other SFI terms do. Even "chaos" seems more metaphoric because it alludes to images of pandemonium. Nouns fall along a continuum between the most abstract, for example, "truth," "justice," "love," on the one side; and the most concrete, "rock," "water," "bone" on the other. "Complexity" is one of those terms at the far abstract end of the pole.

Why, then, does "complexity" serve to define a new science if it is so vague? We have seen many SFI members come to this question. It seems to function as an all-embracing concept precisely because it defies specificity. Consider any of the terms associated with paradigm shifts and you find them to be mostly abstract: the Renaissance, the Age of Enlightenment, the Neo-classical Era, the Age of the Romantics, the Industrial Revolution. The terms that serve to define meta-theories do not do so well, however, in creating new insights within the confines of those theories. It would make no sense to say, "Our lives are futile, a constant Industrial Revolution." But it would make sense to say, "Our lives are futile, the constant turning of gears." The latter metaphor would reveal new knowledge that the process of industrialization is in some ways dehumanizing, while on the other hand, that the process of living is in some ways repetitive and mechanical.

Words that define traditional sciences also are abstract: physics, biology, chemistry, and economics. None of these terms could be

experienced sensually the way you could experience a sand pile catastrophe, a fitness landscape, a rule, or even the edge of chaos.
"Complexity" is no more tangible or no more a metaphor than is
"physics." Both describe an abstract classification for sciences that deal
in very real phenomena. These phenomena often are captured in
theory with the help of concrete metaphors.

The term "complexity" resonates in key with any concept it is
associated with; human society is complex, the solar system is complex, molecular structures are complex, the brain is complex, artificial
worlds on a computer are complex, a transmission with a lot of data is
complex, a transmission with a little data is complex, an individual
organism is complex, the environment it lives in is complex. None of
these associations seems discordant. By contrast, "fitness landscape,"
as we have seen, always implies some kind of life; this harmonic rings
loudly and reveals the discord in using the term to describe a purely
material phenomenon in physics. When these concrete terms are used
across fields, discordant harmonics are always present, always inviting
debate among scientists.

When scientists debate whether some system is complex or not,
they are not debating whether the harmonics sound pleasing. They
are debating a predicate relationship—literally, whether a particular
system has the qualities attributable to a general adjective. Does X fit
into the larger category, Y? They can choose to delimit that predicate
relationship by adding modifiers, such as Gell-Mann's use of "effective," or by specifying conditions, such as Casti's criteria. These criteria, we recall, state that a complex, adaptable system has a medium
number of intelligent, adaptive agents that share information locally.
In this sense, scientists are coarse graining the meta-science, choosing
which power lens to place on the microscope.

Abstract terms allow users to select the degree of specificity they
will hold those terms to. We can stipulate that we will be considering
a narrow view of complexity, perhaps by adhering to Casti's two-page
discussion of its rules—or to a wide view, perhaps by following
Holland's and Kauffman's looser definitions that use just two words.
We have that choice with abstract, non-metaphoric terms. We cannot,
however, stipulate away the associations that come with concrete
metaphors; we must accept those associations and permit them to
challenge—even modify—our theory. A "fitness landscape" will
always conjure up images of anthropomorphic forms moving about
on a fixed geographical terrain.

Abstract terms also spawn harmonics, but they never sound discordant when heard next to another abstract term. "Life is complex;" "life is "simple;" "love is sorrow;" "love is joy." None of these pairings is discordant. "Atoms are playing a game," is discordant. The scientist hearing that pairing can accept the harmonics of agency and incorporate the discord into his concept of what sounds correct, or he can reject the metaphor because it sounds bad. He cannot ignore it.

"Complexity" works to describe a new meta-science because it sounds good in any branch of the member sciences. But it also is so abstract that it has little theoretical power. Pepper, the philosopher who argued that all theories reduce to a few root metaphors, warns that excessive abstraction can render a theory impotent. He could very well use "complexity" as an example when he writes, "Concepts which have lost contact with their root metaphors are empty abstractions" (113). He continues, "When a world theory grows old and stiff (as periodically it does and then has to be rejuvenated), men begin to take its categories and subcategories for granted and presently forget where in fact these come from, and assume that these have some intrinsic and ultimate cosmic value in themselves" (113). The charge that complexity science is flaky or even mystical strikes close to Pepper's use of the words "cosmic value" to describe an abstraction that perhaps has run wild, unfettered by any concrete images.

That is why Gell-Mann's attempt to recover the root words in "complexity" is valuable. As he shows, the word contains the concept of "plaited together." The abstraction becomes concrete in a way that ultra-abstractions like "truth" or "love" probably never were. The number of folds plaited together is tantamount to the length of the rule describing the phenomena. Here the most interesting situations are those that are folded, but not so much as to be tangled. The image of folding, of course, creates harmonics that must be reconciled with the overall theory. Can we say, for example, that the economy has many folds when the image of folds suggests something like a rope, which exists at one point in time? The economy clearly functions over time. These are questions that emerge when a concrete term is excavated from an abstract term, allowing the abstraction once again to function metaphorically.

Gell-Mann has acted almost as an archaeologist of metaphor, digging through the abstraction to find the concrete. While the term "complexity" is too abstract to be a true metaphor, its foundational referent, "fold," is clearly metaphoric; it immediately conjures up harmonic associations of "tangled." A folded or coiled rope, for example, compresses a lot of useful rope into a space for easy storage, but a tangled rope is useless. So perhaps we learn from the metaphoric roots of "complexity" that the rules underlying reality are folded, but not tangled. In a sense, then, the root word "fold" is the metaphoric vehicle for the tenor "complexity."

We have seen in this chapter that the "term" complexity has a long history; it has described phenomena, for example, in classical drama, Enlightenment astronomy, nineteenth century psychology, twentieth century cybernetics, and various philosophies and social sciences. The term has evolved to embrace the meta-science that now bears its name. In nearly every setting, with the possible exception of chemistry, the word "complexity" has functioned abstractly, leaving open many questions. Who is to tell when the plot of a Greek drama becomes complex? How do we differentiate complex celestial motions from simple ones? What makes an idea simple or complex? Efforts to remove the vagueness of "complexity" by adding adjectives are valuable to a point. The power of such a term, however, lies in its persistence and its vagueness. Metaphors and their harmonics subsumed under a grand theory of complexity are necessary to render concrete those vague intuitions.

7

Managing Metaphor Harmonics and Other Challenges of Making Knowledge in Science

It has been some forty years since Thomas Kuhn proclaimed that scientific revolutions are led by paradigm shifts in theory. Kuhn's insight permeated the literature of sociologists, rhetoricians, philosophers, and other scholars of science. The concept of paradigms has been around long enough to spawn its own parodies; the comic strip "Dilbert," for example, has poked fun at the self-importance of so-called "paradigm shift business plans." A newspaper editor I once worked for banned the word from any staff meetings—an example of how a word can be overused to the point of annoyance. As we have seen at the Santa Fe Institute, scientists themselves are wary of making bold claims on human intellectual progress. They are rightfully wary of paradigm shifts.

Although caution is always appropriate whenever academic buzzwords start appearing everywhere, it is clear that Santa Fe Institute scientists and their fellow travelers are up to something new. They are weaving connections among disciplines, developing computer simulations to test those connections, and holding symposia to report on their results. Some of these theories of connections deserve to endure for their novel perspectives—chemistry as syntax, physics as a game of probabilities among particles, information underlying material reality. Other theories that may seem compatible with SFI science are so fanciful or abstractly philosophical as to be meaningless. Presenting the earth as a symbiotic organism, mother "Gaia," is appropriately reverent, perhaps, but far too anthropomorphic and untestable to be of much scientific value.

Separating real paradigm shifts in science theory from spurious ones is the work of scientific discourse. When theory emerges from a think tank, often the only way to test it is to talk about it. Metaphors are irreplaceable in this kind of think-tank discourse. Santa Fe Institute scientists want to understand metaphor because they recognize its power in constructing knowledge. Even researchers like physicist Cosma Rohilla Shalizi, who voice skepticism about the ultimate power of metaphor in science, use it in ways that reveal a great deal of respect. By contemplating how metaphor functions in science and how it differs from other rhetorical devices, scientists involved in theory building become better adept at using language to further their interdisciplinary research. They become adept at using language to strip away the flab of a theory, leaving it taut and potent, but also resilient enough to thrive amid the paradoxes that inevitably resonate with all language.

By working with scientists, technical writers develop confidence that their insights are important, revealing ways in which language can and cannot represent facts in the world. An argument that has been whispered throughout this book is that scientists need technical writers trained in rhetoric to help them negotiate the complexities of language and knowledge. The field of technical writing, or technical communication, is relatively young as an academic discipline. Theorists in this field are struggling to define its boundaries, struggling to find its place among other established disciplines in the academy. If technical writers get nothing else from this book, they should come away with the conviction that they have plenty to offer scientists in the theory building stages. This means not merely in giving Bernadette Longo's coin economic value, but in forging the metal. Following Longo's analogy, we can say that technical writing is rhetoric harnessed to turn imperfect lodes of reality into gems of knowledge. If scientists see mathematics and method as the two Ms that form the basis of scientific assaying, language scholars can rightfully add a third M—metaphor.

Some of the most interesting questions in science are those that consider epistemology, asking what constitutes knowledge and what we call that knowledge. Technical writers have every right to deal head on with these difficult questions in science, what they may mean intrinsically, and what they means to human beings living in the world. Science implies a Platonic order, even if we can never know it. Language determines how we can understand this order. Of course, we must never forget that the subject matter of science is what sets it apart from other human discourse institutions. Technical writers trying to advance new relationships with scientists obviously cannot do so while standing behind sweeping claims that reality is nothing but discourse. C.P. Snow's war between the two cultures of science and literary theory will move closer to peace when literary thinkers back off from their scorched earth policy of claiming all science to be "rhetoric without remainder." Although this claim was useful at one time for its shock value, writers of science know they are writing about real phenomena and things that science has assayed through empirical research, mathematical conjecture, and, yes, discourse. Hence, technical writers can mediate the peace process; they have a stake in both camps.

SPEAK CAREFULLY AND CARRY A PARADIGM SHIFT

None of the rhetorical challenges we have seen are unique to scientists or technical writers at places like the Santa Fe Institute. The challenges begin at the linguistic level, where individual words are called on to make meaning. This leads to the text level, where science writers in real and implied dialogue with their audience score a larger composition of theory using those words deemed meaningful. This, of course, is a social process—language at the pragmatic level—where text engages the community of affiliated researchers and, ultimately, society at large. We can group this array of rhetorical challenges into four related categories:

- The dominant challenge that has filled most of this book is the challenge of semantics, that is, of word meaning. Words obviously do not provide fixed representations of reality; instead they resonate with harmonic associations that provide meaningful impressions of reality. Scientists and other writers of science must manage these harmonics. They can draw upon powerful mathematics to assist in theory building, but they cannot escape the power and paradoxes of language.

- Scientists must present their ideas in an accessible and pleasing writing style, but must not allow the truths inherent in those ideas to be drowned out by the eloquence.

- Writers of science must identify an audience early in the writing process, preferably at the theory building stage. Scientists and technical writers working with scientists invariably struggle when shaping their work both for specialist and generalist audiences; this is best accomplished by revisiting the theories from a new perspective, in essence, by "reinventing" the argument for each audience.

- The final challenge, related to that of audience, is the problem of incommensurability. Scientists from various fields who attempt to communicate with each other through meetings, journal articles, and similar discourse venues find that crossing disciplinary boundaries is never easy. The effort, however, always is worthwhile.

Much of the interview discussion and follow-up analysis continually reverted to problems of meaning: What is a "rule"? What constitutes a "complex" system? Is something a "landscape" if the agents moving about on that landscape are part of it? How is a "pattern" in biology different from a "pattern" in physics? Does the notion of "equilibrium" imply goodness, or is it a value-free term? These questions of meaning inhere at the word level; they are as much a problem of linguistics as of rhetoric. What makes them the subject of a rhetorical study is the awareness that choice of individual words for scientists or anyone else involves consideration of the purpose of a message and its audience. Lawyers, for example, split hairs over word meaning all the time, knowing full well that their ability to persuade a judge and jury is determined by the individual words used in a case. SFI scientists know that when they use a word like "rule" in a colloquium they are inviting debate.

An instrumentalist view of science, found in the post-positivist methodology as it evolved in light of Popper's critique, holds that underlying truth is impossible to prove. Therefore, scientists try to get close enough to the truth to develop theories that work, theories that offer predictive power. SFI scientists also come out of this positivist tradition, although it seems clear that the kind of work they are doing requires them to hope for something more—ideas that are true and not just functional. Think-tank science is intensely philosophical, intensely classical. SFI scientists who hike in the mountains surrounding their center while discussing theory are following the Greek *peripatetic* tradition of theorizing while walking about. These are scientific realists who behave as if the essence of reality can be known, not merely approximated for its predictive power. I would argue that no one at the Santa Fe Institute is searching for theories that merely are not wrong; they want theories that are right and encompassing. Despite their allegiance to the seeming empirical rigor of hypothesis testing, these scientists are not yet producing many falsifiable claims in the traditional sense. Hence, the struggle to define terminology for complexity scientists would seem to be more intense than at centers where applications are the primary research goal.

Words matter a great deal in Platonic philosophy. Plato's dialogues are lengthy discourses about semantics: What are knowledge, beauty, virtue, and so forth? If Plato and his mentor, Socrates, were alive today, they would be welcome at the Santa Fe Institute, an updated "thinkery." Of course, a few scientists would feign suspicion of the philosophers, perhaps harboring lingering feelings of unease that real science should not be so discursive. The tension between post positivism and Platonism is strong throughout science, especially when that science veers farther afield from management of empirical data towards contemplation, dialogue, and simulation.

To say that this kind of think-tank science is Platonic is not to claim that the results of its work could ever lead to one true representation of each aspect of reality. If Platonic philosophy conveys the implication that distinct unchanging Forms lie behind the fuzziness of appearance (and I am not sure that it does), that implication does not hold up in light of science today. Even at their theoretical core, phenomena such as light and gravity exhibit various essences that can seem contradictory. Metaphoric representations of these phenomena, with their harmonic overtones, accurately capture the unsettled nature of reality. The power of metaphors is that they oscillate with probable meanings in much the same way that the referent phenomena do. Metaphoric science is pluralistic; it allows for differing representations to coexist. Helen Longino, the philosopher of science we met earlier in this book, argues that scientific pluralism mediates between theorists who see science as a cognitive process searching for accurate representations and those who see it as a social process searching for acceptable ones. Pluralistic science, Longino argues, can never eliminate the contradictions of language (201).

Plato, of course, was also suspicious of language as the tool of philosophers. Rhetoric in Plato's view was too easy; it diverted attention from the truth. This fundamental problem of rhetoric, the problem of eloquence, is the second rhetorical challenge found at the Santa Fe Institute. It transcends individual word meaning and brings attention to the sentence level and beyond—to the text level. SFI scientists struggle to produce prose that can illuminate nature, but without distorting the image. This is an impossible challenge, of course, because any observation changes that which is observed. Plato's search notwithstanding, we know there can be no "true" representation of reality. All representations are impressionistic.

Nonetheless, scientists and those who write science must strive for good representations, knowing that the goal of perfection is unreachable. For such striving sharpens the thought process. A realistic goal should be to create impressions as carefully as possible, to minimize the impact of observers on the version of reality under study. To use yet another metaphor, nature lovers should walk through the forest carefully on the path, rather than stomping through the bushes.

Many SFI scientists in their interviews suggest that they would like to leave as little trace of their presence as possible, but even a cursory view of SFI texts reveals more than the occasional footprint. Certainly these scientists enjoy the prestige that has come to them because of their association with a trendy new field of science. A few footprints are not bad, for as Plato concluded, a rhetor must love his ideas and want to lead people with them. Scientists like Stuart Kauffman and Murray Gell-Mann clearly love their ideas. They seem to believe that science can offer spiritual succor by revealing the wonder of life's diversity.

Eloquence is a balancing act, perhaps especially so at the places like the Santa Fe Institute because these ideas are so philosophical. In a time when the old philosophies and religions are struggling for direction, many would want these new philosophical ideas—this new path—to be true. Research on the human genome, stem cells, and other twenty-first century scientific explorations is closing in on what it means to be human. As we contemplate the researchers' answers, we also will need the philosophers of complexity science, enlightened theologians, and others to help us cope. Society has always expected guidance from science. Expectations naturally are greater of those scientists who set out to tie together reality's loose ends, such as how the behavior of ants and earthquakes and economies might somehow relate. SFI science clearly has large ambitions, so it understandable that society would have large expectations.

Often scientists answer these expectations in popular books that attempt to generalize their findings and show broad applications for human beings in the world. Santa Fe Institute scientists and others involved in complexity theory and related research have published widely in the popular press; a cursory scan of the science section at large urban bookstores will turn up works by several of the scientists we have followed in this book. Yet, the challenge of reconfiguring research for a popular audience is never trivial. Scientists may be trapped by the Platonic assumption that scientific truth is independent of the observer and, by extension, of the observer's audience. It is a difficult assumption to let go of even with Heisenberg's uncertainty principle always lurking in the back of a scientist's mind.

Yet, it is through implied and real dialogues with specific audiences that a scientist trains her eyes to see different manifestations of reality. The prose of a scientist is most innovative and tractable when it emerges from the mind of the scientist who has an audience in mind from the beginning of the writing process. Scientists are most effective when they return to the invention phase of writing every time they hope to reach a different audience from which the ideas were first conceived. Technical writers trained in rhetoric—with their relentless attention to the twin pillars of rhetorical theory, audience and purpose—can serve a scientist well by also keeping the scientist's mind trained on these pillars.

The fourth rhetorical challenge at the Institute, incommensurability, is a challenge of communicating across disciplines. This is related to the challenge of audience in that it involves one person in a relationship with others for the purpose of developing meaningful knowledge about reality. Yet, incommensurability challenges often are more immediate and tangible. The scientist has to discuss his idea with others, not with an imagined audience. He has to make work in physics seem relevant to computer scientists or biologists, which means translating the discourse conventions of physicists into those of the other disciplines. A technical writer can be of help here by pointing out the terms that a scientist uses almost automatically because they are part of her discipline-specific lexicon and worldview.

By consistently asking a scientist to define terms, a technical writer can help that scientist know which concepts may not be commensurable to other specialists without translation.

Within the Institute, scientists from various fields must learn a language common to them all—one reason that debate over word meaning is so paramount. When biologists and physicists try to share the concept of a "prisoner's dilemma" they are inviting the epistemological and rhetorical challenge of isolating inherent meanings, but also the purely rhetorical challenge of coming to a consensus about which meaning they will agree to use amongst themselves. This is messy work. The only way to surmount problems of incommensurability is to constantly talk them out. The initial debate we saw over the meaning of "rules" took a lot of time, but it was unavoidable and necessary. Even within disciplines people have trouble talking to each other, but the problem is more acute at a place that exists in order to transcend field-dependent thinking. Certainly, at each step along the way toward theory building across disciplines, scientists must stop and ask each other what they mean.

I am aware that in offering this snapshot of evolving theory at a think-tank, I may be skewered by philosophers and scientists who are tired of being told that the problems of science are all linguistic problems. No doubt I will be taken to task for prolonging the moribund academic fiddle-faddle of "talking about talk." Oxford philosopher Bryan Magee delivers what might seem to be a devastating blow to my rhetorical study of science. Citing Popper's attack on positivism, Magee argues that natural scientists produce a lot of useful information without spiraling off into irresolvable debates over what constitutes an "observation," "measurement," "light," "mass," "a number," and so forth (48). Nothing is more humbling than spending several years developing an argument, only to find while approaching the end that a seeming rebuttal already is in paperback at popular bookstores.

Magee is right, as was Popper, but this does not mean that rhetorical problems of science are irrelevant to scientists. This research emphatically shows that debate over meaning is at the core of scientific research. Of course, scientists are pragmatists at heart; problems of language do not stop them from developing life-transforming applications of their research. Perhaps science as science is not about language, but is about an empirical method that is robust enough to endure a little rhetorical ambiguity. Still, science does not arrive fully formed. As Magee points out, again citing Popper, science emerges from various pre-scientific stages. Popper argued, according to Magee, "that as a matter of historical fact all science had emerged by gradual steps out of non-science, out of what logical positivists dismissed as 'metaphysics . . . '" (53). If critics want to argue that metaphor functions at a pre-scientific level—at a metaphysical level—I have no quarrel. Human thought is a complex process that involves multiple actions, some rote, some rational, some intuitive. Perhaps metaphoric thinking is a metaphysical activity rather than a scientific one. The point is that metaphor and all rhetorical methods are essential components in the web of processes that constitute scientific knowledge. Metaphor is alive with harmonics; therefore, it is an indispensable catalyst of scientific thought.

All four rhetorical challenges at the Institute are heightened because the ideas under study there are so new. An etymology of the word "complexity" shows that the multiple meanings associated with that term since Aristotle's time have coalesced into a new science, which considers how order emerges from among the interaction of agents in the world. Traces of the old meanings are evident, but the word clearly has new connotations, as yet still vague, but certainly visible. SFI administrators especially cringe at the idea that theirs is a paradigm shift, but it seems obvious that they are in the midst, if not the forefront, of a significant change in scientific perspective. Of course, it is too early to tell whether the change will mark a significant turn in the course of scientific thinking or merely the illusion of significance. The difference between significance and illusion is told in the applications of science and in ways it which it configures human understanding. The technology of human society functions differently as a result of our knowledge of genetics; the philosophy of society—the way its members understand themselves—is forever changed with the awareness of genetics. Complexity theory has not yet had such a sweeping impact.

But it might do so, extending the meaning and implication of complexity theory well beyond science think-tanks into every facet of society. Perhaps the most significant claim is that information has become the new foundation, the new materiality of science replacing the old physical materiality. Of course, to some extent this is hyperbole, as anyone who bumps her head on a real material tree branch will realize. Even if reality has not changed as much as we think, our perception of it has changed dramatically. We recognize now that information is exchanged among agents leading to behavior among those agents that is more meaningful than the sum of the individual behaviors. We still use gasoline to power steel and plastic automobiles, but we now can perceive those basic components of transportation differently. Gasoline is the remnant of a complex order that emerged millions of years ago in the form of carboniferous plants and prehistoric animal life forms. These forms decayed, lost some information, gained more entropy—but not so much as to return to a random state. These fossil fuels retain information inherent in strings of carbon atoms, which are useful as a source of energy. Matter dissolves, but a message remains for some time after.

We see atoms now not so much as building blocks, but as carriers of information about possible states of reality that cohere into meaningful but temporary structure. Molecular chemistry, says the SFI's Walter Fontana, is "plain syntactical manipulation." The human genome is a string of data that interacts remarkably to give rise to living, intelligent, meaningful beings. Those beings interact to give rise to a complex economy and social structure. Entropy always threatens to overwhelm these structures and meanings with noise, of course, but meaning always finds a new outlet, a new way to percolate forth from the senselessness. Why this happens, why data organizes to become meaningful—this is the unanswered big question underlying the most interesting questions left in science. These are the same questions that gave birth to philosophy 2,500 years ago: What is consciousness? What is free will? What does it mean to be a being in the world? What is knowledge?

These ideas are not simply theoretical, however. Nor are they unique to the Santa Fe Institute. The supremacy of information has been the story of the late twentieth century; the Santa Fe Institute is just a small and recent part of this story. Information exchange has been transforming our real world since the days of the telegraph, when data-rich pulses first traveled across wires in the form of dots and dashes. Many new innovations in telecommunications and computer applications are premised on ideas of cybernetics, which depend entirely on the successful transmission of information broken down into binary units. We take for granted the concept of "memory" in computer science and electronics, but it is an amazing concept. Electrical charges in silicon build up in a way that allows my computer to "remember" the words I have just typed long enough for me to decide whether I want to keep them. Magnetized material in a compact disc aligns in patterns that reproduce a four-minute Beatles song or a four-hour Wagnerian opera on my stereo. We see continual advances—efforts now even to store information in single molecules.

Word meaning is particularly in flux during times like these of paradigm upheaval. Kuhn describes the attempts of normal science to fall in line with a new theory as "empirical work undertaken to articulate the paradigm theory" (27). Such articulation of a theory is what is happening now at the Santa Fe Institute, in part through the debate over word meaning. Kuhn makes it clear that this articulation process is the difficult work that must lie ahead at the Institute because "the search for rules (is) both more difficult and less satisfying than the search for paradigms" (*Structure* 43). Determining that "complexity" would define a new science was the easy part. The difficult work now is in answering how this happens and why it matters; this is the "so what?" question.

The information-as-materiality paradigm is still evolving. Therefore, the Santa Fe Institute cluster of metaphors has not converged yet to a semantically coherent group; the metaphors are mixed. Individual metaphors like "fitness landscape" fairly pulsate with meanings that often contradict each other, implying at the same time the replication of individual agents, perhaps genes; the reproductive success of organisms; the quality of life of those organisms; the relationship of each gene or organism to another; the relationship of each gene to the organism or each organism to the environment; and the constant changes in that environment.

A metaphor like "game theory" at first implies agents with intentionality, but then also the involuntary response of atoms to other atoms. The second implication then modifies the first, suggesting that intentionality even among conscious entities is more involuntary than not. Metaphors "interanimate," as Richards says, or "interact," to use Black's term. Feedback is present. Perhaps these metaphors themselves function at the edge of chaos. The words may settle down to a few fundamental meanings, softening some of the harmonics, when the new paradigm also has settled down. Then it will be time for yet another paradigm.

Paradigms are abstractions. You could not touch the Renaissance even if you were living through it. Scientists using evocative terminology to build theory, even if their proclaimed intent is merely to provide temporary scaffolding, should be aware of whether they are building a collective abstraction or attempting to define it. Abstract terms like "complexity" serve to encapsulate a large body of loosely related ideas. SFI scientists may be frustrated with the term "complexity," for example, when they try to make it function as a metaphor—as a tool for constituting theory—before reigning in its abstractions. As an abstraction, "complexity" contains many related theories, but it builds none. It never has, even before the paradigm shift, which is why it has found a place in so many disciplines for so long. It is a highly adaptable term because it never becomes trapped in specifics. In a way, "complexity" has always served as a proxy for as yet unseen layers of meaning.

If scientists accept the term as an abstraction and use it as such, as a lens through which to view reality according to a general paradigm, they will not be disappointed when it fails to function concretely. They can change the magnifying power of the lens by adding modifiers or by stipulating specific definitions if they want to coarse grain away aspects of that reality. Alternatively, they can excavate the concrete roots from abstract words like "complexity" to make the concepts metaphoric.

Concrete terms generate harmonics that transport meaning across terms. These force the scientists either to modify or reject the original theory so as to accommodate the harmonics. In most post-paradigm shift uses, calling a branch of science "complex" does little more than suggest that it should be studied alongside other systems similarly labeled. Developing concrete metaphoric statements and the accompanying mathematics to say how each system is complex is where the real theory building occurs.

The distinction between theory-constituting metaphors and literary metaphors seems to be arbitrary—making interesting debate for rhetoricians, but otherwise offering little help to rhetors. Language can never simply convey information, for language is information. Language conveys images and narratives. Without it we are left with total abstractions, such as the mathematical representation of a ten-dimensional universe. These abstractions cannot be understood unless they are clothed in language.

Einstein's general relativity theory would be a different theory if it were called something like "spatial-temporal synthesis," even if the mathematics were exactly the same. There is something about the word "relativity" in Einstein's argument that entails more meaning than the sum of the ideas that are subsumed under that term. SFI metaphoric concepts such as "molecules as functions," "epsilon machines," and "signaling" inspire concrete images that radiate with other images, or resonate with harmonics. Concrete metaphors help to chisel meaning out of the abstractions even when those metaphors pulsate with multiple meanings. That is the paradox of a metaphor: It builds knowledge even when its own meanings remain unsettled.

By accommodating the mixed harmonics of a metaphor such as "signaling," for example, biologists are able to build a theory of complex communication among animal life. They can develop computer simulations of agents who selectively send information to each other following certain stimuli (a stink bug giving off an odor when attacked) and they can develop simulations of agents that are imbued with unchanging, differentiating traits (a yellow bee). Thus, they can refine signaling theory in this way to allow for these differences in agent-based sporadic signaling and species-based constant signaling, but only after recognizing that the metaphor implies both kinds. They can accommodate the harmonics of essential metaphors by harnessing these tones within other rhetorical devices, such as *antimetabole.*

Metaphors can lose their potency in science, as may be the case with "game theory." Arguably, it is simply a term that stands for a mathematical method—at least that is the way scientists intend for it to function. But the harmonics do not go away just because scientists have learned to tune them out. In society, "game theory" resonates with disturbing implications of human beings calculating their behavior absent the moral considerations that help blend extremes of black and white into gray. We should be thankful the metaphor retains its overtones. The implications of game theory should not be abstracted out or hidden in mathematical matrices. Scientists should continue to use game theory to develop other theories, but always being aware of its implications.

Recognizing that metaphors are powerful in science, we might argue that scientists should spend their days brainstorming for new ones. But certainly metaphors cannot be called up as needed any more than evidence can be produced on command to support a hypothesis. A new metaphor follows the need for one. Rhetoric is impelled by exigency; rhetoric absent a need sounds hollow and empty. So, for example, I developed my metaphoric statement about the sun as a furnace and used it throughout the early chapters to exhibit how meaning moves across the tenor and vehicle of a metaphor. The statement worked fine and conjured up some meaningful images, although it certainly built no new scientific theory. As we saw, however, it is conceivable that such an image could have been useful for constituting theory in the past. The reason the sun-as-furnace metaphor did not build new theory now is because I was not impelled to develop any theory. I had no scientific exigency. I was simply looking for a plausible example. The metaphor did not come about as a result of my trying to explain reality, as it perhaps could have come to an ancient human contemplating the baked earth of a dried riverbed.

For contrast, let us close this section by considering the running metaphor throughout this project, that of metaphor as a musical note. This note could be sounded on any instrument, such as the string of a violin or the column of air of a clarinet. From that image emerges the concept of harmonics that we have seen to describe the multiple meanings inherent in metaphor. This musical image came to me because I was groping for a way of envisioning the power and problems of metaphor. I needed a theory. A brief analysis of this metaphor will reveal why it has been useful in developing this book. Arguably, the book exists because I searched for a metaphor in order to make a cognitive leap. The larger goal here is to drive home the point that metaphor is a powerful tool in constituting theory because it always makes possible such leaps.

The metaphor of words-as-musical-harmonics endures scrutiny and, for that matter, is enriched by our knowing that harmonics in music can both "color" the sound of instruments and render some instruments discordant in certain circumstances. Whether a scientist appreciates the harmonics of a metaphor depends in part on the piece of music (the theory) he is trying to "play" and the instruments (the words and empirical techniques) he is using. A scientist using a "prisoner's dilemma" metaphor to theorize the behavior of automata in a computer simulation might appreciate the anthropomorphic overtones if he is deliberately trying to blur the notion of intentionality. The goal of discord in music is to challenge established sensibilities— a goal also found in good rhetorical science.

Of course, the running metaphor in this book generates its own associations that may sound discordant in context of metaphor theory. Metaphors, like all words, evolve and change meanings over time. Musical harmonics from a piano key do not change frequencies from one time the key is struck until another, unless the piano goes out of tune. Yet, the way we appreciate a cluster of notes changes over time as our musical sensibilities change. So it seems my running metaphor can be accommodated; it retains its sonority. It is not true, of course, but it resonates tunefully amid today's epistemologies of rhetorical and linguistic theory and the philosophy of science. That is all one could ask of any abstract idea.

Harmonics are present in electronic theory, too, where they almost always are unwanted because they create spurious signals from the main frequency. These spurious signals can cause interference along a radio band, for example. So metaphor harmonics behaving as spurious radio frequency signals would always be a problem. A scientist using "prisoner's dilemma" as a primary "frequency" (a metaphorically derived method of research) who is not trying to be philosophical might want to filter out spurious "frequencies" (associations of agency) prior to applying the theory to magnetized atoms. When seeking results for an empirical study, spurious philosophical overtones are interference.

Coincidentally, the term "harmonics" also is related to the concept of degrees of freedom in mechanical engineering—returning us to the information-as-the-new-materiality argument. Engineers speak of "harmonic motion" as the periodic motion of a system, such as when a piston shaft rotates in an automobile engine. Such a system is simple and has only one degree of freedom, since the position of the piston can always be expressed by one number—its distance from the cylinder head (Den Hartog 23). Other systems like a vibrating string have multiple degrees of freedom. A lot of information must be known to describe the position of any section of that string at any time, in part because of harmonic waves. Multiple harmonics are full of information. Presumably when the information is meaningful, we get the complex timbre of a Stradivarius violin. When the meaning is lost in excessive unorganized information, we get the seemingly random sound of cacophonous bells or bass notes on the piano sounded together.

Science centers like the Santa Fe Institute are organizing information coming from all scientific directions into meaningful sounds, and are adding to human knowledge. Metaphor plays a fundamental role in that research. Such scientists who respond to the harmonics of their ideas by further research and contemplation are doing what scientists have been doing since the dawn of intellectual understanding. Of course, metaphors and their harmonics can distort a scientist's image of the truth or imply a greater sense of scientific certainty than the results of rigorous scientific exploration would support. Scientists and technical writers assisting will accept this paradox when they, like centuries of scientists before them, recognize that the benefits of metaphor far outweigh the risks.

Works Cited

Alexander, Margaret. Personal interview. 6 July 1999.

Aristotle. "Categories." Trans. E. M. Edghill. *The Works of Aristotle: Vol. 1.* Chicago: Encyclopaedia Britannica, 1952. 5–21.

———. *Poetics.* Trans. Kenneth A. Telford. Chicago: Gateway Edition, 1970.

———. *The Art of Rhetoric.* Trans. H.C. Lawson-Tancred. London: Penguin, 1991.

Arrow, Kenneth. Telephone interview. 30 August 1999.

Asimov, Isaac. *The Intelligent Man's Guide to the Biological Sciences.* New York: Basic Books, 1960.

Baake, Kenneth. "Chaos Theory in Economics and Production." Thesis. U Texas, El Paso, 1995.

———. "What Is the Santa Fe Institute?" *Santa Fe Institute Home Page.* Santa Fe, NM: Santa Fe Institute, 1998. URL: http//www.santafe.edu.

———. "Swarm on the Move." *SFI Bulletin* 13:1 (1998): 18–22.

———. "Inside SFI: Language Issues." *SFI Bulletin* 14:2 (1999): 28–29.

———. "Scientific Models at the Santa Fe Institute." Unpublished article.

———. "Initial Impressions." Unpublished report.

Bacon, Francis. "Novum Organum: Book I." *Man and the Universe: The Philosophers of Science.* Eds. Saxe Commins and Robert N. Linscott. New York: Pocket Books, 1954. 73–58.

Bak, Per. *How Nature Works: The Science of Self-Organized Criticality.* New York: Springer-Verlag, 1996.

Ballati, Susan. Personal interview. 15 October 1999.

Barker, Andrew. *Scientific Method in Ptolemy's Harmonics*. Cambridge: Cambridge UP, 2000.

Barth, Robert J., ed. *Religious Perspectives in Faulkner's Fiction: Yokunapatawpha and Beyond*. Notre Dame: U of Notre Dame P, 1972.

Basolo, Fred, and Ralph G. Pearson. *Mechanisms of Inorganic Reactions: A Study of Metal Complexes in Solution*. 2nd ed. 1958. New York: John Wiley and Sons, 1968.

Bazerman, Charles. "Reporting the Experiment: The Changing Account of Scientific Doings in the Philosophical Transactions of the Royal Society, 1665–1800." *Landmark Essays on Rhetoric of Science Case Studies*. Ed. Randy Allen Harris. Mahwah, NJ: Lawrence Erlbaum Associates, 1997. 169–186.

Benson, Philippa Jane. "Changing Moorings in Scientific Writing: Suggestions to Authors, Allusions for Teachers." *Essays in the Study of Scientific Discourse: Methods, Practice, Pedagogy*. Ed. John T. Battalio, Stamford, CT: Ablex, 1998. 209–225.

Bentham, Jeremy. *An Introduction to the Principles of Morals and Legislation*. 1780. London: Oxford at the Clarendon P, 1907.

Bereiter, Carl, and Marlene Scardamalia. *The Psychology of Written Composition*. Hillsdale, NJ: Lawrence Erlbaum and Associates, 1987.

Bertalanffy, Ludwig von. *General System Theory*. New York: George Braziller, 1969.

Best, Steven, and Douglas Kellner. "Postmodern Science." *The Postmodern Turn*. New York: The Guilford Press, 1997.

Black, Max. *Models and Metaphors*. Ithaca, NY: Cornell UP, 1962.

———. "More About Metaphor." *Metaphor and Thought*. Ed. Andrew Ortony. Cambridge, UK: Cambridge UP, 1979. 19–43.

Blackburn, Simon. Ed. *The Oxford Dictionary of Philosophy*. Oxford: Oxford UP, 1996.

Boyd, Richard. "Metaphor and Theory Change: What is 'Metaphor' a Metaphor for?" *Metaphor and Thought*. Ed. Andrew Ortony. Cambridge, UK: Cambridge UP, 1979. 356–408.

Bradbury, Jack W., and Sandra L. Vehrencamp. *Principles of Animal Communication*. Sunderland, MA: Sinauer Associates, 1998.

Brockman, John. *The Third Culture*. New York: Touchstone, 1995.

Brown, James Robert. *Philosophy of Mathematics: An Introduction to the World of Proofs and Pictures*. London and New York: Routledge, 1999.

Byrne, David. *Complexity Theory and the Social Sciences: An Introduction*. New York: Routledge, 1998.

Campbell, John Angus. "Charles Darwin: Rhetorician of Science." *Landmark Essays on Rhetoric of Science Case Studies*. Ed. Randy Allen Harris. Mahwah, NJ: Lawrence Erlbaum, 1997. 3–17.

Capra, Fritjof. *The Web of Life*. New York: Anchor Books, 1996.

Casti, John L. *Would-Be Worlds: How Simulation Is Changing the Frontiers of Science*. New York: John Wiley and Sons, 1996.

———. Personal interview. 14 October 1999.

Cicero. *De Inventione. De Optimo; Genere Oratorum; Topica*. Trans. H.M. Hubbell. Cambridge: Cambridge UP (Loeb Classical Library), 1949.

Cohen, I. Bernard. *Science and the Founding Fathers: Science in the Political Thought of Thomas Jefferson, Benjamin Franklin, John Adams and James Madison*. New York: W.W. Norton, 1997.

Colyvan, Mark. *The Indispensability of Mathematics*. New York: Oxford UP, 2001.

"Complex." *The Oxford English Dictionary*. 2nd ed. 1989.

"Complexion." *The Oxford English Dictionary*. 2nd ed. 1989.

"Complexity." *The Oxford English Dictionary*. 2nd ed. 1989.

"Complicate." *The Oxford English Dictionary*. 2nd ed. 1989.

Comte, Auguste. "The Positive Philosophy." *Mind & the Universe: The Philosophers of Science*. Eds. Saxe Commins and Robert N. Linscott. New York: Pocket Books, 1954. 223–241.

Connor, Kathleen, and Nathan Kogan. "Topic-Vehicle Relations in Metaphor: The Issue of Asymmetry." *Cognition and Figurative Language*. Eds. Richard P. Honeck and Robert R. Hoffman. Hillsdale, NJ: Lawrence Erlbaum Associates, 1980. 283–308.

Copernicus, Nicholas. "On the Revolution of Celestial Spheres." *Man and the Universe: The Philosophers of Science*. Eds. Saxe Commins and Robert N. Linscott. New York: Pocket Books, 1954. 45–72.

Cowan, George. Personal interview. 2 July 1999.

Danto, Arthur C. *Connections to the World: The Basic Concepts of Philosophy.* Berkeley: U of California P., 1989.

Darwin, Charles. *The Origin of Species.* 1859. London: Penguin, 1982.

Darwin, Francis, ed. *The Life and Letters of Charles Darwin.* Vol. II. New York: D. Appleton and Co., 1898.

Davidson, Donald. "What Metaphors Mean." *On Metaphor.* Ed. Sheldon Sacks. Chicago: U of Chicago P, 1978. 29–46.

Dawkins, Richard. *The Selfish Gene.* Oxford: Oxford UP, 1989.

De Man, Paul. "The Epistemology of Metaphor." *On Metaphor.* Ed. Sheldon Sacks. Chicago: U of Chicago P, 1978. 11–28.

Den Hartog, J. P. *Mechanical Vibrations.* 1935. New York: Dover, 1985.

Dennett, Daniel C. *Darwin's Dangerous Idea: Evolution and the Meanings of Life.* New York: Touchstone, 1995.

Derrida, Jacques. "White Mythology: Metaphor in the Text of Philosophy." *Margins of Philosophy.* Trans. Alan Bass. Chicago: U of Chicago P, 1982. 207–271.

Descartes, Rene: *Discourse on Method and Meditations.* Trans. Laurence J. Lafleur. Indianapolis: Liberal Arts P, 1960.

Doyle, Richard. *One Beyond Living: Rhetorical Transformations of the Life Sciences.* Stanford: Stanford UP, 1997.

Dulle, Suzanne. Personal interview. 7 July 1999.

Edmonds, David and John Eidinow. *Wittgenstein's Poker: The Story of a Ten-Minute Argument Between Two Great Philosophers.* New York: HarperCollins, 2001.

"Equilibrium." *The Oxford English Dictionary.* 1933.

Fahnestock, Jeanne. "Accommodating Science: The Rhetorical Life of Scientific Facts." *The Literature of Science: Perspectives on Popular Scientific Writing.* Ed. Murdo William McRae, Athens, GA: U of Georgia P, 1993. 17–36.

———. "Arguing in Different Forums: The Bering Crossover Controversy. *Landmark Essays on Rhetoric of Science Case Studies.* Ed Randy Allen Harris. Mahwah, NJ: Lawrence Erlbaum Associates, 1997.

———. *Rhetorical Figures in Science.* New York: Oxford UP, 1999.

Feyerabend, Paul. *Against Method.* New York and London: Verso, 1988.

Fleming, David. Personal communication. Spring, 1998.

Fontana, Walter. Personal interview. 6 July 1999.

———. Telephone interview. 19 August 1999.

Foss, Sonja K., Karen A. Foss, and Robert Trapp. "Richard Weaver."
    Contemporary Perspectives on Rhetoric. 2nd ed. Prospect Heights, Ill:
    Waveland P, 1985. 55–86.

Foucault, Michel. *The Order of Things: An Archaeology of the Human
    Sciences.* 1971. New York: Random House, 1994.

Gaonkar, Dilip Parameshwar. "The Idea of Rhetoric in the Rhetoric of
    Science." *Rhetorical Hermeneutics: Invention and Interpretation in the
    Age of Science.* Eds. Alan G. Gross and William M. Keith. Albany,
    NY: State University of New York P, 1997. 25–85.

Gell-Mann, Murray. *The Quark and the Jaguar: Adventures in the Simple
    and the Complex.* New York: W.H. Freeman, 1994.

Gentner, Dedre, and Michael Jeziorski. "The Shift From Metaphor to
    Analogy in Western Science." *Metaphor and Thought.* 2nd ed. Ed.
    Andrew Ortony. Cambridge, UK: Cambridge UP, 1993. 447–480.

Gilder, George. *Microcosm: The Quantum Revolution in Economics and
    Technology.* New York: Simon & Schuster, 1989.

Goldberg, Ellen. Personal interview. 8 July 1999.

Gross, Alan G., and William M. Keith. "Introduction." *Rhetorical
    Hermeneutics: Invention and Interpretation in the Age of Science.* Eds.
    Alan G. Gross and William M. Keith. Albany, NY: State University
    of New York P, 1997. 1–22.

Gross, D(avid). "Opening Questions." *Unified String Theories.* Eds.
    M(ichael) Green and D(avid) Gross. Singapore: World Scientific
    Publishing Co., 1986. 1–2.

Hall, Calvin S., and Vernon J. Nordby. *A Primer of Jungian Psychology.* New
    York: Signet, 1973.

Hampel, Clifford A. *Glossary of Chemical Terms.* New York: Van Nostrand
    Reinhold Co., 1976.

Haraway, Donna J. *Crystals, Fabrics, and Fields: Metaphors of Organicism in
    Twentieth-Century Developmental Biology.* New Haven, CT: Yale UP,
    1976.

"Harmonics." *The Norton/Grove Concise Encyclopedia of Music*. Ed. Stanley Sadie. New York: W.W. Norton, 1994.

Harris, Richard J., Mary Anne Lahey, and Faith Marsalek. "Metaphors and Images: Rating, Reporting, and Remembering." *Cognition and Figurative Language*. Eds. Richard P. Honeck and Robert R. Hoffman. Hillsdale, NJ: Lawrence Erlbaum Associates, 1980. 163–181.

Heilbron, J.L. *Electricity in the 17th and 18th Centuries: A Study in Early Modern Physics*. 1979. Mineola, NY: Dover Publications, 1999.

Helmreich, Stefan. *Silicon Second Nature: Culturing Artificial Life in a Digital World*. Berkeley: U of California P, 1998.

Hely, Tim. Personal interview. 13 October 1999.

Henig, Robin Marantz. *A Dancing Matrix: Voyages Along the Viral Frontier*. New York: Alfred A. Knopf, 1993.

Hoffman, Robert R. "Metaphor in Science." *Cognition and Figurative Language*. Eds. Richard P. Honeck and Robert R. Hoffman. Hillsdale, NJ: Lawrence Erlbaum Associates, 1980. 393–423.

Holland, John. *Hidden Order: How Adaptation Builds Complexity*. Reading, MA: Addison-Wesley, 1995.

Horgan, John. *The End of Science*. New York: Broadway Books, 1997.

Jarrett, Dennis. "World Class Science Happens in Santa Fe." *Santa Fe Reporter*. Sept. 17–23, 1997: 11+.

Jeans, Sir James. *Physics and Philosophy*. 1943. New York: Dover Publications, 1981.

Jen, Erica. Various conversations, telephone and personal. 1997–1999.

Johnson, George. *Fire in the Mind: Science, Faith, and the Search for Order*. New York: Alfred A. Knopf, 1995.

———. *Strange Beauty: Murray Gell-Mann and the Revolution in Twentieth Century Physics*. New York: Alfred A. Knopf, 1999.

———. Telephone interview. 22 November 1999.

Johnson, Michael G., and Robert Malgady. "Toward a Perceptual Theory of Metaphoric Comprehension." *Cognition and Figurative Language*. Eds. Richard P. Honeck and Robert R. Hoffman. Hillsdale, NJ: Lawrence Erlbaum Associates, 1980. 259–282.

Joshi, Shareen. Personal interview. 8 July 1999.

Kant, Immanuel. *Critique of Pure Reason*. Trans. Norman Kemp Smith (1929). New York: St. Martin's P, 1965.

———. *The Philosophy of Kant*. Ed. Carl J. Friedrich. New York: The Modern Library, 1949.

Kauffman, Stuart. *The Origins of Order: Self-Organization and Selection in Evolution*. New York: Oxford UP, 1993.

———. *At Home in the Universe*. New York: Oxford UP, 1995.

———. Personal interview. 6 August 1999.

———. Telephone interview. 8 February 2001.

Keiger, Dale. "Why Metaphors Matter." *Johns Hopkins Magazine*. 19.1 (1998): 40–45.

Keller, Evelyn Fox. *Secrets of Life, Secrets of Death: Essays on Science and Culture*. New York: Routledge, 1992.

Kelly, Susanne, and Mary Ann Allison. *The Complexity Advantage: How the Science of Complexity Can Help Your Business Achieve Peak Performance*. New York: McGraw-Hill, 1999.

Kiel, L. Douglas. *Managing Chaos and Complexity in Government: A New Paradigm for Managing Change, Innovation, and Organizational Renewal*. San Francisco: Jossey-Bass, 1994.

King, Lesley. Various telephone conversations. 1999–2000.

Krinsky, Fred. *Democracy and Complexity: Who Governs the Governors?* Beverly Hills, CA: The Glencoe P., 1968.

Kuhn, Thomas S. *The Structure of Scientific Revolutions*. 3rd ed. Chicago: U of Chicago P, 1996.

———. "Metaphor in Science." *Metaphor and Thought*. Ed. Andrew Ortony. Cambridge, UK: Cambridge UP, 1979. 409–419.

Lachmann, Michael. Personal interview. 1 July 1999.

Lachmann, Michael, and Carl T. Bergstrom. "When Honest Signals Must Be Costly." Santa Fe Institute working paper 99-08-059. 1999.

Lake, Anthony. "Managing Complexity in U.S. Foreign Policy." Imprint of a speech before the World Affairs Council of Northern California, March 14, 1978. In *Washington Department of State, Bureau of Public Affairs, Office of Public Communication*, newsletter, 1978.

Lakoff, George, and Mark Johnson. *Metaphors We Live By*. Chicago: U of Chicago P, 1980.

Lanham, Richard. *A Handlist of Rhetorical Terms.* 2nd Ed. Berkeley, U of California P, 1991.

Latour, Bruno, and Steve Woolgar. *Laboratory Life: The Construction of Scientific Facts.* Princeton, NJ: Princeton UP, 1986.

LeFevre, Karen Burke. *Invention as a Social Act.* Carbondale, IL: Southern Illinois UP, 1987.

Leibniz, Gottfried Wilhelm. "The Principles of Nature and of Grace, Based on Reason." 1714. *Leibniz Selections.* Ed. Philip P. Wiener. New York: Charles Scribner's Sons, 1951. 522–533.

Lincoln, Yvonna S., and Egon G. Guba. *Naturalistic Inquiry.* Beverly Hills: SAGE, 1985.

Locke, David. *Science as Writing.* New Haven, CT: Yale UP, 1992.

Locke, John. *An Essay Concerning Human Understanding.* Ed. Peter H. Nidditch. London: Oxford UP, 1975.

Longino, Helen. *The Fate of Knowledge.* Princeton: Princeton UP, 2002.

Longo, Bernadette. *Spurious Coin: A History of Science, Management, and Technical Writing.* Albany, NY: State U of New York P, 2000.

Lucretius. *The Way Things Are.* Trans. Rolfe Humphries. Bloomington, IN: Indiana UP, 1969.

Magee, Bryan. *Confessions of a Philosopher: A Personal Journey Through Western Philosophy, From Plato to Popper.* New York: The Modern Library, 1999.

Mathiesen, Thomas. *Apollo's Lyre: Greek Music and Music Theory in Antiquity and the Middle Ages.* Lincoln, Neb: U of Nebraska P., 1999.

McCloskey, Donald (now Deirdre). *The Rhetoric of Economics.* Madison, WI: U of Wisconsin P, 1985.

Mehra, Jagdish. " 'The Golden Age of Theoretical Physics': P.A.M. Dirac's Scientific Work from 1924 to 1933." *Aspects of Quantum Theory.* Eds. Abdus Salam and E.P. Wigner. London: Cambridge U. P., 1972. 17–59.

"Metaphor." *Webster's New Collegiate Dictionary.* 1977.

Minor, Dennis E. "Albert Einstein on Writing." *Journal of Technical Writing and Communication.* 14.1 (1984): 13–18.

Montgomery, Scott. *The Scientific Voice*. New York: The Guilford Press, 1996.

Moore, Chris. Personal interview. 2 July 1999.

Morowitz, Harold. "What's in a Name?" *Complexity*. 1.4 (1995-1996): 7–8.

———. "Metaphysics, Metaphor, Meta-metaphor, and Magic." *Complexity*. 3.4 (1998): 19–20.

Myers, Greg. "Texts as Knowledge Claims: The Social Construction of Two Biology Articles." *Landmark Essays on Rhetoric of Science Case Studies*. Ed. Randy Allen Harris. Mahwah, NJ: Lawrence Erlbaum Associates, 1997. 187–215.

"Neural Networks." *Encyclopedia of Computer Science*. 3rd ed. Eds. Anthony Ralston and Edwin D. Reilly. New York: Van Nostrand Reinhold, 1993. 929–930.

Ong, Walter J. "The Writer's Audience is Always a Fiction." *Cross-Talk in Comp Theory: A Reader*. Ed. Victor Villanueva Jr. Urbana, IL: NCTE, 1997. 55–76.

Ortony, Andrew. "Metaphor: A Multidimensional Problem." *Metaphor and Thought*. Ed. Andrew Ortony. Cambridge, UK: Cambridge UP, 1979. 1–16.

———. "Some Psycholinguistic Aspects of Metaphor." *Cognition and Figurative Language*. Eds. Richard P. Honeck and Robert R. Hoffman. Hillsdale, NJ: Lawrence Erlbaum Associates, 1980. 69–83.

Ostrom, Elinor. Personal interview. 7 July 1999.

Papineau, David. "Philosophy of Science." *The Blackwell Companion to Philosophy*. Eds. Nicholas Bunnin and E.P. Tsui-James. Oxford: Blackwell Publishers, 1996. 290–324.

Pauling, Linus, and Roger Hayward. *The Architecture of Molecules*. San Francisco: W.H. Freeman, 1964.

Pepper, Stephen C. *World Hypotheses*. 1942. Berkeley: U of California P, 1970.

Perelman, Ch., and L. Olbrechts-Tyteca. *The New Rhetoric: A Treatise on Argumentation*. 1958. Trans. John Wilkinson and Purcell Weaver. Notre Dame, IN: U of Notre Dame P, 1971.

Peters, F.E. *Greek Philosophical Terms: A Historical Lexicon*. New York: New York UP, 1967.

Phillips, Derek. *Wittgenstein and Scientific Knowledge: A Sociological Perspective.* Totowa, NJ: Rowman and Littlefield, 1977.

Pitts, Mary Ellen. "Reflective Scientists and the Critique of the Mechanistic Metaphor." *The Literature of Science: Perspectives on Popular Scientific Writing.* Ed. Murdo William McRae, Athens, GA: U of Georgia P, 1983. 249–272.

Plato. *Plato's Complete Works.* Ed. John M. Cooper. Indianapolis, IN: Hacket Publishing Co., 1997.

Polanyi, M. *The Tacit Dimension.* Garden City, N.Y: Doubleday, 1966.

Popper, Karl. "Science: Conjectures and Refutations." *Conjectures and Refutations: The Growth of Scientific Knowledge.* 1962. New York: Harper, 1968. 33–65.

Pullman, Bernard. ed. *The Emergence of Complexity in Mathematics, Physics, Chemistry, and Biology: Proceedings, Plenary Session of the Pontifical Academy of Sciences,* 27–31 October 1992. Princeton: Princeton UP, 1996.

Purrington, Robert. *Physics in the Nineteenth Century.* New Brunswick, NJ: Rutgers UP, 1997.

Pylyshyn, Zenon. W. "Metaphorical Imprecision and the 'Top Down' Research Strategy." *Metaphor and Thought.* Ed. Andrew Ortony. Cambridge, UK: Cambridge UP, 1979. 420–436.

Randall, John Herman Jr. *The Making of the Modern Mind.* 50th Anniversary ed. New York: Columbia UP, 1976.

Reichenbach, Hans. *The Rise of Scientific Philosophy.* Berkeley: U of California P, 1951.

"Rhetorica ad Herennium: Book IV." *The Rhetorical Tradition: Readings From Classical Times to the Present.* Eds. Patricia Bizzell and Bruce Herzberg. Boston: Bedford Books, 1990. 252–292.

Rice, Thomas Jackson. *Joyce, Chaos, and Complexity.* Urbana, Ill: U of Illinois P, 1997.

Richards, I.A. *The Philosophy of Rhetoric.* New York: Oxford UP, 1965.

Richardson, Ginger. Various conversations, personal, telephone, and email, 1997–2000.

Ricoeur, Paul. *The Rule of Metaphor: Multi-disciplinary Studies of the Creation of Meaning in Language.* Trans. Robert Czerny et al. Toronto: U of Toronto P, 1997.

Rorty, Richard. "Science as Solidarity." *The Rhetoric of the Human Sciences: Language and Argument in Scholarship and Public Affairs*. Eds. John S. Nelson, Allan Megill and Donald N. McCloskey. Madison, WI: U of Wisconsin P, 1987.

Seashore, Carl. *Psychology of Music*. 1938. New York, Dover Publications, 1967.

Secord, James A. *Victorian Sensation: The Extraordinary Publication, Reception, and Secret Authorship of Vestiges of the Natural History of Creation*. Chicago: U of Chicago P, 2000.

Shalizi, Cosma Rohilla. "Scientific Models: Claiming and Validating." *SFI Bulletin*. 13:2 (1998): 8-12.

———. Personal interview. 15 October 1999.

Shalizi, Cosma Rohilla, and James P. Crutchfield. *Computational Mechanics: Pattern and Prediction, Structure and Simplicity*. Santa Fe Institute working paper 99-07-044. Santa Fe Institute, 1999.

Shapin, Steven. *The Scientific Revolution*. Chicago: U of Chicago P, 1996.

Sheehan, Richard D. Johnson. "Metaphor as Hermeneutic." *Rhetoric Society Quarterly*. 29.2 (1999): 47–64.

Siegfried, Tom. *The Bit and the Pendulum: How the New Physics of Information Is Revolutionizing Science*. New York. John Wiley, 2000.

Snow, C.P. *The Two Cultures and the Scientific Revolution*. New York: Cambridge UP, 1959.

Solow, Robert. "How Economic Ideas Turn to Mush." *The Spread of Economic Ideas*. Eds. David Colander and A.W. Coats. Cambridge, UK: Cambridge UP, 1989. 75–84.

Spiegel, Henry W. *The Growth of Economic Thought*. 3rd ed. Durham, NC: Duke UP, 1991.

Streufert, Siegfried and Robert W. Swezey. *Complexity, Managers, and Organizations*. Orlando, FL: Academic P, 1986.

Timm, John Arrend. *General Chemistry*. 4th ed. New York: McGraw-Hill, 1966.

Toulmin, Stephen. *The Uses of Argument*. 1958. London: Cambridge UP, 1969.

Turner, Mark. *Reading Minds: The Study of English in the Age of Cognitive Science*. Princeton: Princeton UP, 1991.

Turner, Roger. Personal interview. 8 July 1999.

Verbrugge, Robert. "Transformations in Knowing: A Realist View of Metaphor." *Cognition and Figurative Language.* Eds. Richard P. Honeck and Robert R. Hoffman. Hillsdale, NJ: Lawrence Erlbaum Associates, 1980. 87–125.

Waldrop, M. Mitchell. *Complexity: the Emerging Science at the Edge of Order and Chaos.* New York: Touchstone, 1992.

Winsor, Dorothy A. "Invention and Writing in Technical Work: Representing the Object." *Written Communication.* 11 (1994): 227–250.

Witten, Edward et all. "Overview of *K*-Theory Applied to Strings." August 2000. Online. Los Alamos National Laboratory archives. 15 June 2002. URL: http://xxx.lanl.gov/PS_cache/hep-th/pdf/0007/0007175.pdf

Wolfram, Stephen. *Cellular Automata and Complexity: Collected Papers.* Reading, MA: Addison-Wesley Publishing, 1994.

Young, Robert. "Darwin's Metaphor: Does Nature Select?" *Darwin's Metaphor.* Cambridge: Cambridge UP, 1985. 79-82

Index

abiotic (non-living systems), 190
accommodation
  harmonics in music and, 72
  metaphor and, 12, 81, 89, 114, 221
Adams, John, 76
adaptation
  as a requirement for complexity, 195
adapting, as a scientific phenomenon, 34, 42
agency, 123. *See also* intentionality
alchemy, 46, 48–49
Alexander, Margaret, 98, 100, 103
  on eloquence and style, 154–55
algorithm, 127
algorithmic information content, 194–95
allele, 166
Allison, Mary Ann, 199
alliteration, 164
Ampere, Andre-Marie, 115
analogue, 77–78
analogy, 76–78
Anasazi. *See* Native Americans
Anderson, Philip, 23, 36
ant colony and computer model, 129–30,
  133, 213
anthropomorphism in science, 126, 128, 143,
  207
antimetabole, 155–56, 219
aposiopesis, 117
apparent non-stationarity, 174
archaeology, 37
argument theory, 172
Aristotelian
  metaphysics, 54
  philosophy, 49
Aristotle, 13, 22, 28, 34, 38, 49, 197
  complexity and, 180
  essences of matter, 87

invention and, 169
persuasion and, 149, 156
Poetics, 58
  complex plots in, 183, 201
  science as analytic system and, 46
  on simile and analogy as forms of
    metaphor, 77
The Art of Rhetoric, 77, 156
Aristoxenus, 11, 63
Arrow, Kenneth, 23, 106, 121
  on biological metaphors in economics, 175
  on "complexity" as a negative word, 192
  on eloquence and style, 153–56
  on emergence, 133
  on equilibrium, 140–42
  on limitations of theory, 88–89
  on rhetoric, 96
  on slogans, 100–1
Arthur, Brian, 154
articulation of theory, 217
Artificial Anasazi, 21
artificial life, 20
Asimov, Isaac, 69
astronomy, 184
At Home in the Universe. See Kauffman,
  Stuart: audience and
atoms, 126
  as carriers of information, 216
  in a magnetic field, 124, 144–45
  metaphors of, 69, 71, 77–78, 201
ATP synthetase, 39
audience, 38, 41
  composite, 170
  importance of to rhetoricians, 163
  incommensurability and, 213
  insiders, 26
  lay, 166

audience, *cont'd*
  for *Metaphor and Knowledge*, 4
  need for clear definitions, 159
  for science writing, 151, 160–72
axon, 99
as a wire, 135

Baake
  "Initial Impressions" report, 27, 29
  attempt at insider writing, 37
  call for reinvention in science writing, 167
  interview method, 80
  metaphor choices, 26
  narrative in science writing, 39
  roiling soup metaphor, 35, 42
  theory
    explained. *See* harmonics: metaphor
    possible rebuttal of, 214–15
  writing assignments, 21, 25–42
Babylonians, 45
Bacon, Francis, 5, 22, 47, 49–50, 54–55, 152
Bak, Per, 114, 196
Ballati, Susan, 173, 176–77
Barker, Andrew, 11–12
Barthes, Robert, 198–99
Basolo, Fred, 187
Bazerman, Charles, 50
Beardsley, Monroe, 64
beauty as a complex idea, 184
bee, 127–28, 133, 145
Beethoven, Ludwig van, 9
being and becoming in Platonic philosophy, 63
Benson, Phillipa Jane, 163
Bentham, Jeremy, 184
Bereiter, Carl, 41
Bergstrom, Carl T., 155
Bertalanffy, Ludwig von, 190
Best, Steven, 19
biology, 146, 155, 172, 213–14
  complexity science and, 100, 180
  fitness landscapes and, 132
  game theory and, 143
  intentionality and, 127
  neural networks and, 135–38
  writing and, 164, 169
biotic (living) systems, 190
Black, Max, 38, 202
  frame and focus in metaphor theory, 63, 193–94

metaphor as interaction, 62–64, 144
metaphor as substitution or comparison, 58–59
on models in science, 77–78
Blackburn, Simon, 47
Blake, unidentified author in the *OED*, 186
Bogdanov, Alexander, 190
Bohr, Niels, 71
Boltzmann, Ludwig, 188
Boolean algebra, 104
Borges, Jorge Luis, 115
Boyd, Richard, 114, 125, 137
  on metaphor as theory constitutive or literary, 70–71
  on ostensive representation, 59, 62
Boyle, Robert, 47, 49
Bradbury, Jack, 125
brain
  biology of, 105, 135
  metaphor and, 99
  metaphors for activity of, 136–37
  metaphor of as computer, 70–71, 78, 134–38, 145
brainstorming, 168
Brown, James Robert, 91–92, 109, 111
Burke, Edmund, 184
Burke, Kenneth, 55, 172
Buss, Leo, 112–13, 138, 174
Byrne, David, 198

Campbell, John, 51
Capra, Fritjof, 188–89
Casti, John, 153, 203
  on balancing substance and style, 159
  on complex adaptive systems, 194–96
  on equilibrium, 140–41
  on rules, 129
  on SFI language, 100–105
  on writing to understand, 157
causality, 82, 121
  as a pattern. *See* patterns in physics as opposed to in biology
cause and effect, 40, 49, 109
cell, 128
  brain, 134–36
  cooperation, 144
  in computer models, 126
  metaphor of as factory, 77
cellular automata, 20, 126
cellulose, 187